Fundamental
Principles of
Polymeric Materials

SPE MONOGRAPHS

FUNDAMENTAL PRINCIPLES OF POLYMERIC MATERIALS

STEPHEN L. ROSEN

Professor and Chairman
Department of Chemical Engineering
The University of Toledo
Toledo, Ohio

A WILEY-INTERSCIENCE PUBLICATION

JOHN WILEY & SONS

New York · Chichester · Brisbane · Toronto · Singapore

Library of Congress Cataloging in Publication Data:

Rosen, Stephen L., 1937–
    Fundamental principles of polymeric materials.

    (SPE monographs, ISSN 0195-4288)
    Previously published as: Fundamental principles
of polymeric materials for practicing engineers.
1971.
    "A Wiley-Interscience publication."
    Bibliography: p.
    Includes index.
    1. Polymers and polymerization.   I. Title.
II. Series: SPE monographs.

TA455.P58R63 1981   668.9   81-10320

ISBN 0-471-08704-1   AACR2

Printed and bound in the United States of America by Braun-Brumfield, Inc.

20 19 18 17 16 15 14 13 12

Series Preface

The Society of Plastics Engineers is dedicated to the promotion of scientific and engineering knowledge of plastics and to the initiation and continuation of educational programs for the plastics industry. Publications, both books and periodicals, are major means of promoting this technical knowledge and of providing educational materials.

New books, such as this volume, have been sponsored by the SPE for many years. These books are commissioned by the Society's Technical Volumes Committee, and, most importantly, the final manuscripts are reviewed by the Committee to ensure accuracy of technical content. Members of this Committee are selected for outstanding technical competence and include prominent engineers, scientists, and educators.

In addition, the Society publishes *Plastics Engineering, Polymer Engineering and Science (PE & S), Journal of Vinyl Technology, Polymer Composites,* proceedings of its Annual, National, and Regional Technical Conferences (ANTEC, NATEC, RETEC), and other selected publications. Additional information can be obtained from the Society of Plastics Engineers, 14 Fairfield Drive, Brookfield Center, Connecticut 06805.

v

Preface

This book was written to provide an appreciation of those fundamental principles of polymer science and engineering that are currently of practical relevance. I hope that the reader will obtain both a broad, unified introduction to the subject matter that will be of immediate practical value and a foundation for more advanced study.

The first edition of the book was intended primarily as a self-study guide for practicing engineers and scientists. Despite the fact that it was a well-kept secret, it also achieved modest success as an academic text. In this edition I have included additional material which I hope will make it more suitable as a text. By so doing I sincerely hope that I have not made it any less useful to the original audience. To this end, all the problems are still presented as worked-out examples. I have tried to emphasize a qualitative understanding of the underlying principles before tackling the mathematical details, so that the former may be appreciated independently of the latter (I don't recommend trying it the other way around, however), and I have included practical illustrations of the material whenever possible.

The treatments of gel permeation chromatography, linear viscoelasticity, and what has been termed "polymer reaction engineering" have been expanded considerably. New material has been included on gel formation, the three-dimensional solubility parameter, and molecular weight distributions. There has been a general updating of the material, particularly in the section on technology, although this is like shooting at a rapidly moving target.

Obviously, the choice of material to be covered involves subjective judgment on the part of the author. This, together with space limitations and the rapid expansion of knowledge in the field, has resulted in omission or shallow treatment of many interesting subjects. The references and selected readings have been specifically chosen to aid the reader who wishes to pursue a subject in greater detail.

A word to the student: To derive maximum benefit from the worked-out examples, make an honest effort to answer them *before* looking at the solutions. If you can't do one, you've missed some important points in the preceding material, and you ought to go back over it.

STEPHEN L. ROSEN

Pittsburgh, Pennsylvania
April 1981

Acknowledgments

I'd like to thank my students for proving time and again that the best way to learn is to teach; my teachers, Professors B. Maxwell, L. Rahm, H. Pohl, the late A.V. Tobolsky, and, in particular, Ferdinand Rodriguez, who was also my research advisor, for making the macromolecular gospel so fascinating; and my industrial friends and colleagues for keeping me aware of the important "real world" problems.

My former department head, Tom Fort, gave me a semester off so I could put the manuscript together. My colleagues, Drs. Guy Berry, Ethel Casassa, Hershel Markovitz, and Dennis Prieve were kind enough to read portions of the manuscript and provide helpful suggestions. The manuscript was typed by Mrs. Dolores Dlugokecki, who, despite my handwriting and the ongoing trials and tribulations of the Pirates and Steelers, hardly ever missed a word or symbol. The 06-709 class of 1979 struggled with the rough manuscript and, I hope, caught most of the errors. I'm grateful to them all for their help.

S.L.R.

Contents

*Fundamental
Principles of
Polymeric Materials*

CHAPTER 1 _____

Introduction

Since World War II, polymeric materials have been the fastest growing segment of the United States chemical industry. It has been estimated that more than 25% of the chemical research dollar is spent on polymers, with a correspondingly large proportion of technical personnel working in the area.

A modern automobile contains over 200 lbs (100 kg) of plastics, and this does not include paints, the rubber in tires, or the fibers in tires and upholstery. With the increasing need to save fuel and therefore weight, polymers will continue to replace traditional materials in the automotive industry. Similarly, the applications of polymers in the building construction industry (piping, resilient flooring, siding, thermal and electrical insulation, paints, decorative laminates, etc., etc.) are already impressive and will become even more so in the future. A trip through a supermarket will quickly convince anyone of the importance of polymers in the packaging industry (bottles, films, trays, etc.). Many other examples could be cited, but, to make a long story short, the use of polymers now outstrips that of metals on a volume basis.

Since nearly all modern polymers have their origins in petroleum, it has been argued that this increased reliance on polymers constitutes an unnecessary drain on energy resources. However, the raw materials for polymers account for less than 2% of total petroleum consumption; so while the petroleum shortage will continue to drive the price of polymers up (along with everything else), even the total elimination of synthetic polymers would not contribute significantly to the conservation of hydrocarbon resources. Furthermore, when *total* energy costs (raw materials plus energy to manufacture) are compared, the polymeric item often comes out well ahead of its traditional counterpart, for example, glass versus plastic beverage bottles.

There are five major areas of application for polymers: (1) plastics, (2) rubbers or elastomers, (3) fibers, (4) surface finishes and protective coatings, and (5) adhesives. Despite the fact that the five applications are all based on polymers,

1

and in many cases the same polymer is used in two or more, the industries grew up pretty much separately. It was only after Dr. Herman Staudinger proposed the "macromolecular hypothesis" in the 1920s[1,2] explaining the common molecular makeup of these materials (for which he won the 1953 Nobel Prize in chemistry in belated recognition of the importance of his work) that polymer science began to evolve from the independent technologies. Thus a sound fundamental basis was established for continued technological advances.

Economic considerations alone would be sufficient to justify the impressive scientific and technological efforts expended on polymers in the past several decades. In addition, however, this class of materials possesses many interesting and useful properties that are completely different from those of the more traditional engineering materials and that cannot be explained or handled in design situations by the traditional approaches. A description of three simple experiments should make this obvious.

"Silly putty," a silicone polymer, bounces like rubber when rolled into a ball and dropped. On the other hand, if the ball is placed on a table, it will gradually spread to a puddle. The material behaves like an elastic solid under certain conditions and like a very viscous liquid under others.

If a weight is suspended from a rubber band and the band is then heated (taking care not to burn it), the rubber band will *contract* appreciably. All materials other than polymers will undergo the expected thermal *expansion* upon heating (assuming no phase transformation has occurred over the temperature range).

When a rotating rod is immersed in a molten polymer or a fairly concentrated polymer solution, the liquid will actually climb up the rod. This phenomenon, the Weissenberg effect, is contrary to what is observed with nonpolymer liquids, which develop a parabolic surface profile with the lowest point at the rod as the material is flung outward by centrifugal force.

Although such behavior is unusual in terms of the more familiar materials, it is a perfectly logical consequence of the *molecular structure* of polymers. This molecular structure is the key to an understanding of the science and technology of polymers, and underlies the chapters to follow.

Fig. 1.1 illustrates the questions to be considered:

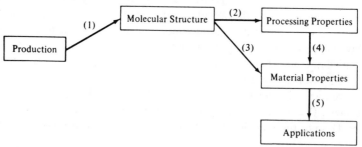

Figure 1.1   The key role of molecular structure in polymer science and technology.

1.   How is the desired molecular structure obtained?

2.   How do the polymer's processing (i.e., formability) properties depend on its molecular structure?

3.   How do its material properties (mechanical, chemical, optical, etc.) depend on molecular structure?

4.   How do material properties depend on a polymer's processing history?

5.   How do its applications depend on its material properties?

The word polymer comes from the Greek "many-membered." Strictly speaking, it could be applied to any large molecule that is formed from a relatively large number of smaller units or "mers," a sodium chloride crystal, for example, but it is most commonly (and exclusively, here) restricted to materials in which the "mers" are held together by covalent bonding, that is, shared electrons. For our purposes, only a few bond valences need be remembered:

$$-\overset{|}{\underset{|}{C}}- \quad \overset{|}{\underset{/ \backslash}{N}} \quad -O- \quad Cl- \quad F- \quad H- \quad -\overset{|}{\underset{|}{Si}}-$$

It is always a good idea to "count the bonds" in any written structure to make sure they conform to the above. A brief, concise review of organic chemistry from the polymer standpoint is available.[3]

The most important constituents of living organisms, cellulose and proteins, are naturally occurring polymers, but we confine our attention largely to synthetic polymers or to important modifications of natural polymers.

REFERENCES

1.   H. Staudinger, *Bericht*, **53**, 1073 (1920).

2.   H. Staudinger and J. Fritsch, *Helv. Chim. Acta*, **5**, 778 (1922).

3.   P. N. Richardson and R. C. Kierstead, *SPE J.*, **25**, 9, 54 (1969).

PART 1

Polymer Fundamentals

CHAPTER 2

Types of Polymers

The large number of natural and synthetic polymers has been classified in several ways. These are outlined below, and in the process many terms important in polymer science and technology are introduced.

2.1   REACTION TO TEMPERATURE

The earliest distinction between types of polymers was made long before any concrete knowledge of their molecular structure existed. It was a purely phenomenological distinction based on their reaction to heating and cooling.

A.   Thermoplastics

It was noted that certain polymers would soften upon heating and then could be made to flow when a stress was applied. When cooled again, they would reversibly regain their solid or rubbery nature. These polymers are known as *thermoplastics*. By analogy, ice is a thermoplastic.

B.   Thermosets

Other polymers, although they might be heated to the point where they would soften and could be made to flow under stress *once*, would not do so reversibly; that is, heating caused them to undergo a "curing" reaction. Sometimes these materials emerge from the synthesis reaction in a cured state. Further heating of these *thermosetting* polymers ultimately leads only to degradation (as often attested to by the smell of a short-circuited electrical appliance) and not to softening and flow. An egg is thermosetting.

7

Continued heating of thermoplastics also leads ultimately to degradation, but they will generally soften at temperatures below their degradation point.

A classic example of these two categories is natural rubber. Introduced to Europe by Columbus, natural rubber did not achieve any commercial significance for centuries because it was a thermoplastic — articles made of it would become soft and sticky on hot days. It was only when Goodyear discovered the curing reaction that converted the polymer to a thermoset — which he called vulcanization in honor of the Roman god of fire — allowing it to maintain its useful rubber properties at much higher temperatures that rubber began to achieve commercial importance.

2.2   CHEMISTRY OF SYNTHESIS

Pioneering workers in the field of polymer chemistry soon observed that they could produce polymers by two familiar types of organic reactions.

A.   Condensation

Polymers formed from a typical organic condensation reaction, in which a small molecule (most often water) is split out, are known, logically enough, as *condensation polymers.* The common esterification reaction of an organic acid and an organic base (alcohol) illustrates the simple "lasso chemistry" involved:

$$R-O\overbrace{H} + \overbrace{HO}-\overset{\overset{\textstyle O}{\|}}{C}-R' \quad \rightarrow \quad R-O-\overset{\overset{\textstyle O}{\|}}{C}-R' \;\; + H_2O$$

Alcohol +   acid   →   ester

Of course, the ester formed in the preceding reaction is not a polymer because we have only hooked up two small molecules, and the reaction is finished far short of anything that might be considered "many membered."

At this point it is useful to introduce the concept of *functionality. Functionality is the number of bonds a "mer" can form with other "mers" in a reaction.* It is obvious from the example above that each of the reactants is monofunctional, and that reactions between monofunctional "mers" cannot lead to polymers. But now consider what happens if each reactant is *di*functional, allowing it to react at each end.

$$HO-R-OH + HO-\overset{O}{\overset{\|}{C}}-R'-\overset{O}{\overset{\|}{C}}-OH \longrightarrow HO-R-O\overset{O}{\overset{\|}{C}}-R'-\overset{O}{\overset{\|}{C}}-OH + H_2O$$

Di alcohol   Di carboxylic acid

or   or

diol   diacid

The resulting product molecule is still difunctional because its left end can react with another diacid molecule and its right end with another molecule of diol. After each subsequent reaction, the growing molecule is still difunctional and capable of undergoing further growth, leading to a true polymer molecule.

In general, the *polycondensation* of x moles of diol with x moles of a diacid to give a *polyester* is

$$x\, HO-R-OH + x\, HO-\overset{O}{\overset{\|}{C}}-R'-\overset{O}{\overset{\|}{C}}-OH \longrightarrow H-\left[O-R-O-\overset{O}{\overset{\|}{C}}-R'-\overset{O}{\overset{\|}{C}}\right]_x OH + (2x-1)\, H_2O$$

Polyester

In the polyester molecule above, the structure in brackets is the *repeating unit* and is what distinguishes one polymer from another, while the $-(O-\overset{O}{\overset{\|}{C}})-$ linkage characterizes all polyesters; that is, the generalized organic groups R and R' may vary widely (with a consequent variation in the properties of the polymer), but as long as the repeating unit contains the $-(O-\overset{O}{\overset{\|}{C}})-$ linkage, the polymer is a polyester. The quantity x is the *degree of polymerization*, the number of repeating units strung together like identical beads in the polymer chain. It is sometimes also called the *chain length*, but it is a pure number, not a length.

Another functional group that is capable of taking part in a condensation is the ami*ne* ($-NH_2$) group, *one* hydrogen of which reacts with a carboxylic acid group in a manner similar to the alcoholic hydrogen to form a polyami*de* or nylon:

$$x\, \overset{H}{\underset{H}{>}}N\,R\,N\overset{H}{\underset{H}{<}} + x\, HO-\overset{O}{\overset{\|}{C}}-R'-\overset{O}{\overset{\|}{C}}-OH \longrightarrow H-\left[\overset{H}{\underset{}{N}}-R-\overset{H}{\underset{}{N}}-\overset{O}{\overset{\|}{C}}-R'-\overset{O}{\overset{\|}{C}}\right]_x OH + (2x-1)\, H_2O$$

Diamine   +   Diacid $\longrightarrow$ Polyamide or Nylon

The $-\left(\!\!\begin{array}{c}H \\ N\end{array}\!\!-\!\!\begin{array}{c}O \\ C\end{array}\!\!\right)-$ linkage characterizes nylons.

The examples above serve to illustrate that reactants must be at least difunctional if a polymer is to be obtained. Molecules with higher degrees of functionality will also lead to polymers. For example, glycerine

$$H\!-\!\underset{\underset{OH}{|}}{\overset{\overset{H}{|}}{C}}\!-\!\underset{\underset{OH}{|}}{\overset{\overset{H}{|}}{C}}\!-\!\underset{\underset{OH}{|}}{\overset{\overset{H}{|}}{C}}\!-\!H \text{ is}$$

trifunctional in a polyesterification reaction.

In addition to using two monomers with the same functional group at each end, it is also possible to form condensation polymers from a single monomer containing the two complementary reactive groups in the same molecule:

$$\underset{\text{Hydroxy acid}}{\overset{\overset{O}{\overset{\|}{}}}{HO\!-\!R\!-\!C\!-\!OH}} \qquad \underset{\text{Amino acid}}{\overset{\overset{O}{\overset{\|}{}}}{\overset{H}{\underset{H}{>}}N\!-\!R\!-\!C\!-\!OH}}$$

In principle, the hydroxy acid is capable of forming a polyester and the amino acid a polyamide (proteins are poly amino acids). The reactions do not always proceed in a straightforward fashion, however. If R is large enough, say three carbon atoms or more, the difunctional monomers above may "bite their own tails," condensing to form a cyclic structure:

$$\underset{\text{Amino acid}}{\overset{\overset{O}{\overset{\|}{}}}{\overset{H}{\underset{H}{>}}N\!-\!R\!-\!C\!-\!OH}} \longrightarrow \underset{\text{Lactam}}{R\overset{C\overset{\|}{}O}{\underset{N}{<\atop|}{\atop H}}} + H_2O$$

This cyclic compound can then undergo a *ring scission* polymerization, in which the polymer is formed without splitting out a small molecule, which had been eliminated previously in the cyclization step.

$$x \ R\overset{C\overset{\|}{}O}{\underset{N}{<\atop|}{\atop H}} \longrightarrow \underset{\text{Polyamide}}{\left[\!\!\begin{array}{c}H \ \ O \\ N\!-\!R\!-\!C\end{array}\!\!\right]_x}$$

Despite the lack of elimination of a small molecule in the actual polymerization step, the products can be thought of as being formed by a direct condensation from the monomer and are usually considered condensation polymers.

Note that in the polyamide above, the characteristic nylon linkage $-\left(N-C\right)-$ with H on the N and O double-bonded to C, has been split up in the repeating unit as written. This illustrates the somewhat arbitrary location of the brackets but should not obscure the fact that the polymer is a nylon.

B.   Addition

The second polymer-formation reaction is known as addition polymerization and its products as *addition polymers.* Addition polymerizations have two distinct characteristics:

1.   No molecule is split out; hence the repeating unit has the same chemical formula as the monomer,
2.   The polymerization reaction involves the opening of a double bond.*

Monomers of the general type $C{=}C$ undergo addition polymerization:

$$x \ \ C{=}C \ \ \longrightarrow \ \ \left[C{-}C\right]_x$$

The double bond "opens up," forming bonds to other monomers at each end; so *a double bond is difunctional.* The question of what happens at the ends of the polymer molecule will be deferred for a discussion of polymerization mechanisms.

An important subclass of the double-bond containing monomers is the vinyl monomers, $\overset{\text{H X}}{\underset{\text{H H}}{C{=}C}}$. Addition polymerization is occasionally referred to as vinyl polymerization. Table 2.1 lists some commercially important vinyl monomers.

---

*Although aromatic rings are often symbolized by ⬡ , this is a poor representation of a resonance-stabilized structure that is completely inert to addition polymerization. They are more properly symbolized by ⬡ to avoid confusion with the ordinary double bond.

Table 2.1   Some commercially important vinyl monomers

Monomer	$-X$
Ethylene	$-H$
Vinyl chloride	$-Cl$
Styrene	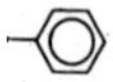 $-\bigcirc$
Propylene	$-CH_3$
Acrylonitrile	$-C{\equiv}N$

Example 1. Lactic acid can be dehydrated to acrylic acid according to the following reaction:

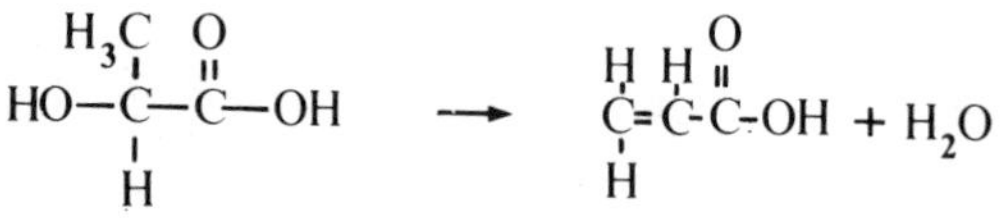

$$
\underset{\text{Lactic acid}}{\overset{\displaystyle H_3C\;\;O}{HO-\underset{\underset{H}{|}}{C}-\overset{\|}{C}-OH}} \longrightarrow \underset{\text{Acrylic acid}}{\overset{\displaystyle \;\;\;\;O}{\underset{\underset{H}{|}}{C}=\overset{H}{C}-\overset{\|}{C}-OH} + H_2O}
$$

Each acid forms a polymer. Write the structural formulas for the repeating unit of each polymer.

Solution.   Lactic acid is a hydroxy acid and will undergo condensation polymerization, splitting out water from the $-OH$ and $-\overset{\displaystyle O}{\overset{\|}{C}}-OH$ groups to give a polyester.

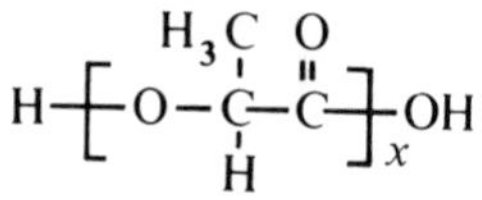

$$
H-\left[\!\!\left[O-\overset{\displaystyle H_3C\;\;O}{\underset{\underset{H}{|}}{C}-\overset{\|}{C}}\right]\!\!\right]_x OH
$$

Note again that in the above formula, the characteristic polyester linkage $-\!\!\left(O-\overset{\displaystyle O}{\overset{\|}{C}}\right)\!\!-$ is split up. Acrylic acid is a vinyl monomer and will undergo addition polymerization to give

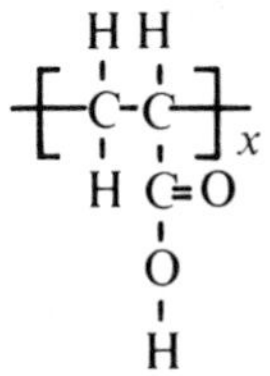

$$\begin{array}{c} \text{H} \ \text{H} \\ \Bigl[\!\!\begin{array}{c} | \ \ | \\ \text{C-C} \\ | \ \ | \\ \text{H} \ \text{C=O} \end{array}\!\!\Bigr]_x \\ | \\ \text{O} \\ | \\ \text{H} \end{array}$$

Dienes (the carbon atoms of which are numbered from one end)

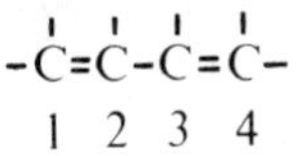

$$-\overset{|}{\text{C}}=\overset{|}{\text{C}}-\overset{|}{\text{C}}=\overset{|}{\text{C}}-$$
$$1 \quad 2 \quad 3 \quad 4$$

are also capable of undergoing addition polymerization. Addition polymerization of dienes results in an *unsaturated* polymer, that is, a chain that contains double bonds. Furthermore, there are several possibilities for the addition reaction. If the monomer is symmetrical with regard to substituent groups, it can undergo 1,2 addition and 1,4 addition. If unsymmetrical, there is the added possibility of 3,4 addition. (For symmetrical dienes the 1,2 and 3,4 reactions are the same.) This is illustrated below for the addition polymerization of isoprene (2-methyl 1,3 butadiene).

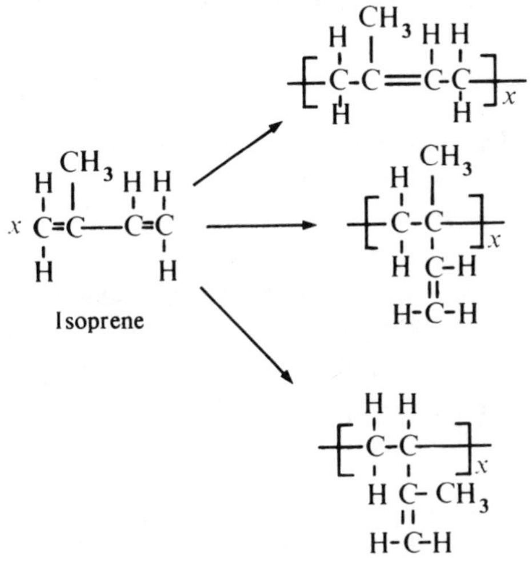

1, 4 Polyisoprene

Isoprene

1, 2 Polyisoprene

3, 4 Polyisoprene

The 1,2 and 3,4 reactions are sometimes known as *vinyl* addition because part of the diene monomer simply acts as an X group in a vinyl monomer.

2.3   STRUCTURE

As an appreciation of the molecular structure of polymers was gained, three major structural categories emerged.

A.   Linear

If a polymer is built from strictly *difunctional* monomers, the result is a *linear* polymer chain. The term linear is somewhat misleading, however, because the molecules are never stretched out in a truly linear fashion. In general, an isolated linear molecule, say in a dilute solution, assumes a more or less random twisted and tangled configuration when not subjected to an external stress. Mathematical treatments of such flexible chains, in fact, are based on the random walk problem in three dimensions: If a drunk is turned loose at a lamp post, with the direction of each of his steps being completely random, where will he be after x steps (if you can imagine him staggering in three dimensions)? In practice, polymer molecules are never isolated, of course, and the many chains in a bulk sample are twisted and tangled together. A common analogy is a bowl of cooked spaghetti. A far better analogy, in view of the constant thermal motion of the chain segments (for those with a strong stomach) is a bowl of wriggling worms. A scale model of a typical molecule, if made from $\frac{1}{2}$-cm-diameter clothesline, would be on the order of 3 m long.

Random Copolymers

Polymers consisting of chains that contain a single repeating unit are known as *homopolymers*. If, however, the chains contain a random arrangement of two separate and distinct repeating units, the polymer is known as a *random copolymer*, or just plain *copolymer*. A random copolymer might be formed by the addition polymerization of a mixture of two different vinyl monomers, A and B (the degree of "randomness" depends on the relative amounts and reactivities of A and B, as will be seen later), and be represented as

AABAAABBABAAB

and called poly(A-co-B), where the first repeating unit listed is the one present in the greater amount. For example, a random synthetic rubber copolymer of 75% butadiene and 25% styrene would be termed poly(butadiene-co-styrene). Of course, ter- and higher multipolymers are possible.

It must be emphasized that the products of condensation polymerizations that require two different monomers to provide the necessary functional groups, for example, a diacid and a diamine, *are not* copolymers, because they contain only one repeating unit. If, however, two different diamines were to be used, leading to two distinct repeating units, the product would be a copolymer.

Example 2. Illustrate the repeating units which result when three moles of hexamethylene diamine (I) are condensed with two moles of adipic acid (II) and one mole of sebacic acid (III), and name the resulting copolymer.

$$H_2N\!-\!\!\left(CH_2\right)_6\!\!-\!NH_2 \qquad HO\!-\!\!\overset{\overset{\text{O}}{\|}}{C}\!-\!\!\left(CH_2\right)_4\!\!-\!\overset{\overset{\text{O}}{\|}}{C}\!-OH \qquad HO\!-\!\!\overset{\overset{\text{O}}{\|}}{C}\!-\!\!\left(CH_2\right)_8\!\!-\!\overset{\overset{\text{O}}{\|}}{C}\!-OH$$

(I)   (II)   (III)

Solution. The two repeating units are

$$\left[\!\!\overset{\overset{\text{H}}{|}}{N}\!-\!\!\left(CH_2\right)_6\!\!-\!\overset{\overset{\text{H}}{|}}{N}\!-\!\overset{\overset{\text{O}}{\|}}{C}\!-\!\!\left(CH_2\right)_4\!\!-\!\overset{\overset{\text{O}}{\|}}{C}\!\!\right] \quad \text{and} \quad \left[\!\!\overset{\overset{\text{H}}{|}}{N}\!-\!\!\left(CH_2\right)_6\!\!-\!\overset{\overset{\text{H}}{|}}{N}\!-\!\overset{\overset{\text{O}}{\|}}{C}\!-\!\!\left(CH_2\right)_8\!\!-\!\overset{\overset{\text{O}}{\|}}{C}\!\!\right]$$

from (I) and (II)   from (I) and (III)

The formal (if unwieldy) name for the copolymer containing these two repeating units is poly(hexamethylene adipamide-co-hexamethylene sebacamide).

Block Copolymers

Under certain conditions it is possible to form linear chains that contain long contiguous blocks of two (or more) repeating units combined in the chains, which is called a *block copolymer*.

AAAAAAAAAAAAAAAAAAAAAAAAAABBBBBBBBBBBBBBBBBBB

Such a polymer would be termed poly(A-b-B).

B. Branched

If a few points of tri (or higher) functionality are introduced (either intentionally or through side reactions) at random points along linear chains, *branched* molecules result. Branching can have a tremendous influence on the properties of polymers through steric (geometric) effects.

Graft Copolymers

Under specialized conditions, branches of repeating unit B may be "grafted" to a backbone of linear A. This structure is known as a *graft copolymer*:

$$
\begin{array}{c}
\text{AAAA----A-----A} \\
\end{array}
$$

The graft copolymer above would be called poly(A-g-B), the backbone repeating unit being the one listed first.

C.   Crosslinked or Network

As the length and frequency of the branches on polymer chains increase, the probability that the branches will finally reach from chain to chain, connecting them together, becomes greater. When all the chains are finally connected together in three dimensions by these *crosslinks*, the entire polymer mass becomes one tremendous molecule, a *crosslinked* or *network* polymer. A bowling ball, for example, has a molecular weight on the order of 10^{28}, all the polymer in it being connected to form one molecule by crosslinks.

Crosslinked or network polymers may be formed in two ways: (1) by starting with reaction masses containing sufficient amounts of tri- or higher functional monomer, or (2) by chemically creating crosslinks between previously formed linear or branched molecules ("curing"). The latter is precisely what vulcanization does to natural rubber, and this fact serves to introduce the connection between the phenomenological "reaction to temperature" classification and the more fundamental concept of molecular structure. This important connection will be clarified through a discussion of bonding in polymers.

Example 3.   Show (a) how a linear, unsaturated polyester is produced from ethylene glycol (I) and maleic anhydride (II), and (b) how the linear, unsaturated polyester is crosslinked with a vinyl monomer, styrene, for example.

$$
\begin{array}{c}
\text{H H} \\
| \ | \\
\text{HO-C-C-OH} \\
| \ | \\
\text{H H}
\end{array}
$$

$$
\begin{array}{c}
\text{H} \diagdown \text{C} - \text{C} \diagup^{\textstyle O} \\
\quad \| \qquad \diagdown O \\
\quad \text{C} - \text{C} \diagdown \\
\text{H} \diagup \qquad^{\textstyle O}
\end{array}
$$

(I)   (II)

Solution.   First, realize that an acid anhydride is simply a diacid with a mole of water split out from the two acid groups (this is the only common example of an acid group condensing with itself; i.e., you cannot ordinarily form polymers this way). When considering the reaction of an acid anhydride, (conceptually) hydrate it back to the diacid:

$$
\begin{array}{c}
\text{H} \diagdown \text{C} - \text{C} \diagup^{\textstyle O} \\
\quad \| \qquad \diagdown O \\
\quad \text{C} - \text{C} \diagdown \\
\text{H} \diagup \qquad^{\textstyle O}
\end{array}
\quad + \text{H}_2\text{O} \quad \longrightarrow \quad
\begin{array}{c}
\text{O} \qquad\ \text{O} \\
\| \qquad\quad \| \\
\text{HO-C-C=C-C-OH} \\
\qquad | \ \ | \\
\qquad \text{H H}
\end{array}
$$

Maleic anhydride   Maleic acid

Then condense the diacid with the diol to form a polyester with one double bond per repeating unit:

$$
\begin{array}{c}
\text{H H} \\
| \ | \\
\text{HO-C-C-OH} \\
| \ | \\
\text{H H}
\end{array}
+
\begin{array}{c}
\text{O} \quad\ \text{O} \\
\| \qquad \| \\
\text{HO-C-C=C-C-OH} \\
\quad | \ | \\
\quad \text{H H}
\end{array}
\rightarrow
\text{H}\!\!\left[\!
\begin{array}{c}
\text{H H} \quad\ \text{O H H O} \\
| \ | \qquad \| \ | \ | \ \| \\
\text{O-C-C-O-C-C=C-C} \\
| \ | \\
\text{H H}
\end{array}
\!\right]_x\!\!\text{OH}
$$

The double bond in the maleic acid is inert toward polycondensation. Note that the degree of unsaturation (average number of double bonds per repeating unit) could be varied from zero to one by employing mixtures of a saturated diacid, phthalic anhydride (III), for example, with the maleic to form copolyesters with saturated and unsaturated repeating units.

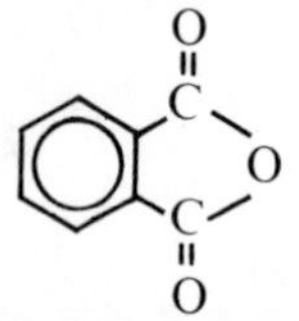

(III)

Commercially, a low degree of polymerization, x, is maintained (say 8–10) (by techniques to be considered in Chapter 9) so that the product is a viscous liquid.

The linear, unsaturated polyester is then diluted with a liquid vinyl monomer, most often styrene. Before use, a chemical that promotes additional polymerization (Chapter 10) is added, causing the vinyl monomer to undergo addition copolymerization with the double bonds in the polyester, forming a highly crosslinked, rigid network:

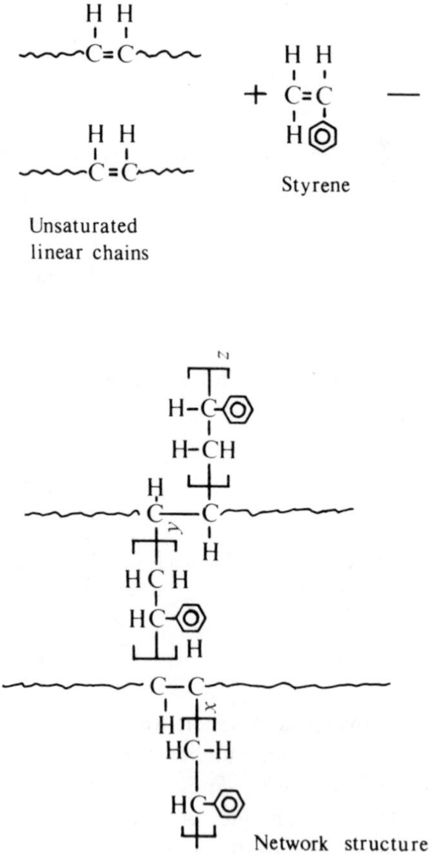

Unsaturated
linear chains

Styrene

Network structure

The liquid polyester–styrene mixture is often used to impregnate fiber glass and is cured to form boat hulls, auto (Corvette) bodies, and other so-called fiber-glass objects (really fiber-glass-reinforced polyester).

Example 4.   Show the structural formulas of the repeating units for each of the following polymers and classify them according to structure and chemistry of formation. All the polymers are commercially important. Most follow the rules outlined above, but some do not and have been included here to illustrate their structures, chemistries of formation, and characteristic linkages.

A.   Polystyrene
B.   Polyethylene
C.   Poly(butylene terephthalate) (PBT)
D.   Poly(ethylene terephthalate) (PET) (Dacron,® Mylar®)
E.   Nylon 6/6 (The numbers designate carbon atoms in the diamine/diacid.)
F.   Nylon 6 (The number designates carbon atoms in the monomer.)
G.   Glyptal (glycerol + phthalic anhydride)
H.   Poly(diallyl phthalate)
I.   Melamine-formaldehyde (Melmac,® Formica®)
J.   Polytetrafluoroethylene (Teflon TFE®)
K.   Poly(phenylene oxide) (PPO) (Hint: polymerized in presence of O_2.)
L.   Polypropylene
M.   Acetal (polyformaldehyde or polyoxymethylene) (Delrin,® Celcon®)
N.   Polycarbonate (Lexan,® Merlon®)
O.   Epoxy or phenoxy
P.   Poly(dimethyl siloxane) (silicone rubber) (Hint: polymerized in presence of H_2O.)
Q.   Polyurethane
R.   Polyimide
S.   Polysulfone (Udel®)

Starting monomers are shown below

A   H–C=CH₂
      Styrene

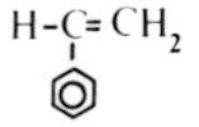

B   H₂C=CH₂
      Ethylene

C   HO–(CH₂)₄–OH
      Butylene glycol
      (1,4 butane diol)

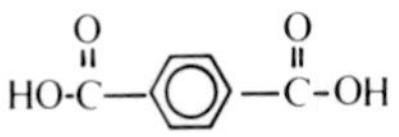

      HO-C(=O)-⟨⟩-C(=O)-OH
      Terephthalic acid

D H_3C-O-$\overset{\text{O}}{\underset{\parallel}{C}}$—⬡—$\overset{\text{O}}{\underset{\parallel}{C}}$-O-$CH_3$

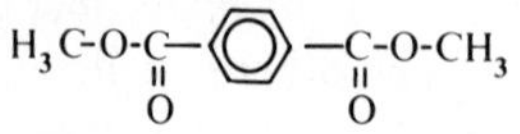

Dimethyl terephthalate

HO-$(CH_2)_2$-OH

Ethylene glycol

E H_2N-$(CH_2)_6$-NH_2

Hexamethylene diamine

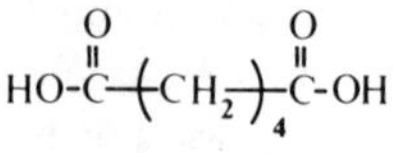

HO-$\overset{\text{O}}{\overset{\parallel}{C}}$-$(CH_2)_4$-$\overset{\text{O}}{\overset{\parallel}{C}}$-OH

Adipic acid

F

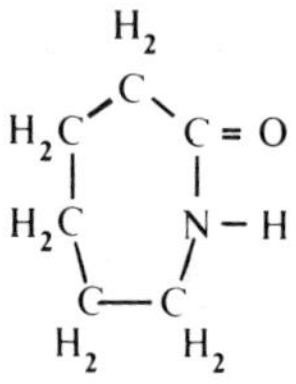

ε–caprolactam

G H_2-C-$\overset{H}{\underset{}{C}}$-C$H_2$ with O O O / H H H

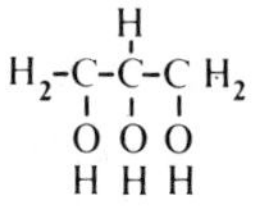

Glycerol

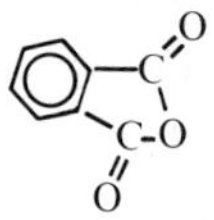

Phthalic anhydride

H

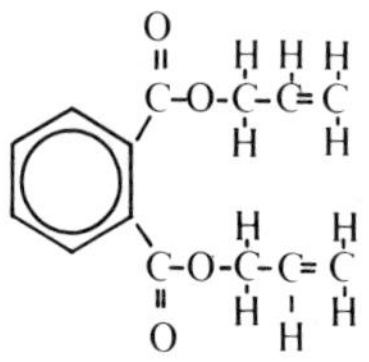

Diallyl phthalate

I   Melamine

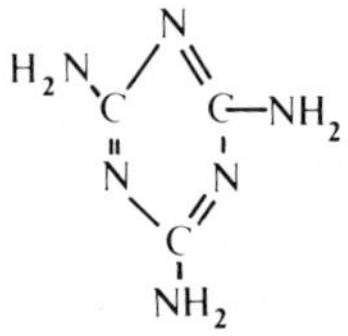

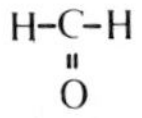

H-C-H
$\overset{\parallel}{O}$

Formaldehyde

J F_2C=CF_2

Tetrafluoroethylene

K

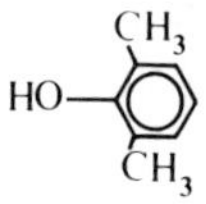

2,6 Dimethyl phenol

L H_2C=$\overset{H}{\underset{CH_3}{C}}$

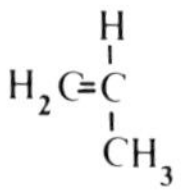

Propylene

M   H-C-H   or   trioxane
$\overset{\parallel}{O}$

Formaldehyde

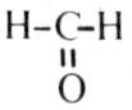

Trioxane

N   HO-⬡-$\overset{CH_3}{\underset{CH_3}{C}}$-⬡-OH

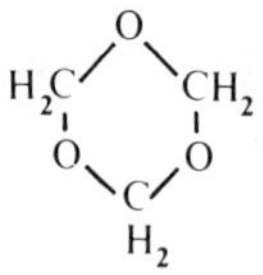

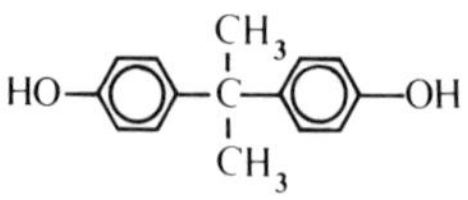

Bisphenol-A

Cl-$\overset{\text{O}}{\underset{\parallel}{C}}$-Cl

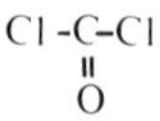

Phosgene

O   HO—⟨benzene⟩—C(CH₃)(CH₃)—⟨benzene⟩—OH

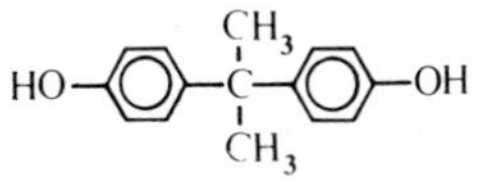

Bisphenol –A

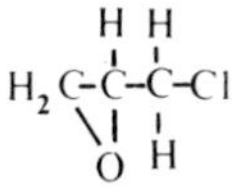

H₂C—C(O)H—C—CH₂—Cl (Epichlorohydrin structure)

Epichlorohydrin

P   Cl–Si(CH₃)(CH₃)–Cl

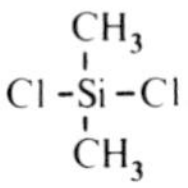

Dimethyl dichlorosilane

Q   HOROH

Diol   or   glycol

O=C=N–R′–N=C=O

Disocyanate

R   (Dianhydride structure: O=C–O–C=O rings joined by R)

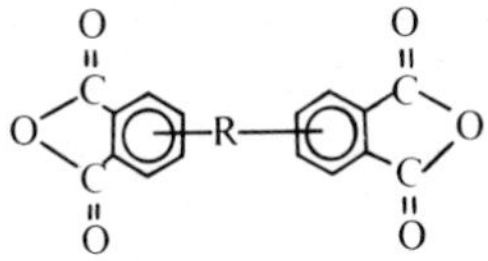

Dianhydride

H₂NR′NH₂

Diamine

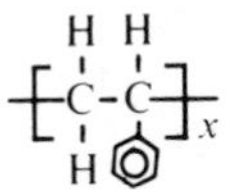

S   NaO—⟨benzene⟩—C(CH₃)(CH₃)—⟨benzene⟩—ONa

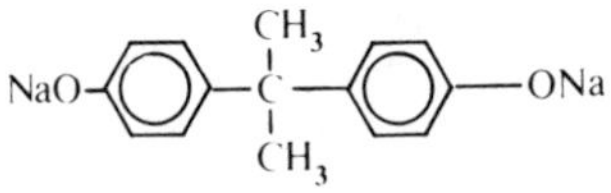

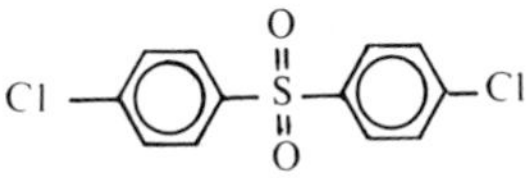

Cl—⟨benzene⟩—S(=O)(=O)—⟨benzene⟩—Cl

Solution

A   (C–C)ₓ with H,H / H, phenyl   Linear addition

B   (C–C)ₓ with H,H / H,H   Linear or branced from side reactions, addition

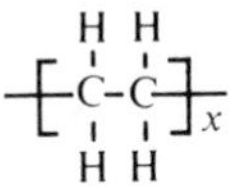

C   H–[O–(CH₂)₄–O–C(=O)–⟨benzene⟩–C(=O)]ₓ–OH   (H₂O out)   Linear. condensation

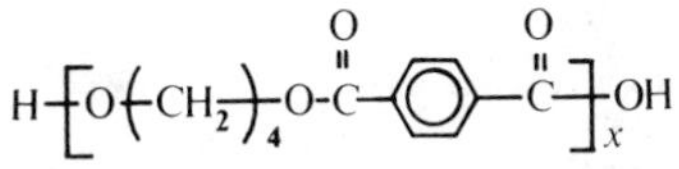

D   H₃C–[O–C(=O)–⟨benzene⟩–C(=O)–O–(CH₂)₂]ₓ–OH   CH₃OH

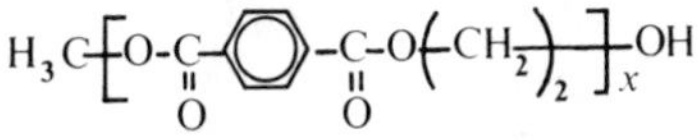

Split out

Linear, condensation

Can also be made from the diacid (see above) splitting out H₂O

E    (H_2O out) linear, condensation

Characteristic nylon linkage

F 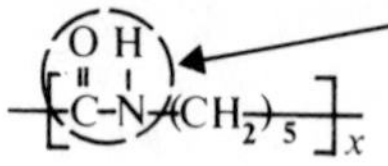   (nothing out)   Monomer made from

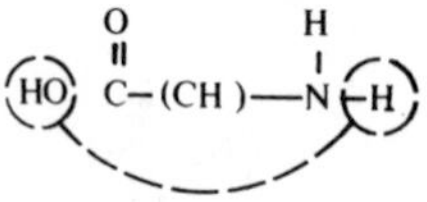

Linear, condensation

G 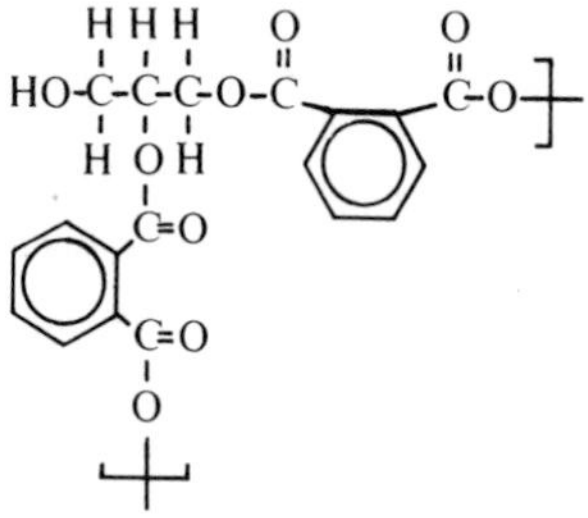   Condensation, structure depends on ratio of reactants (chapter 9), but usually will be highly crosslinked

H

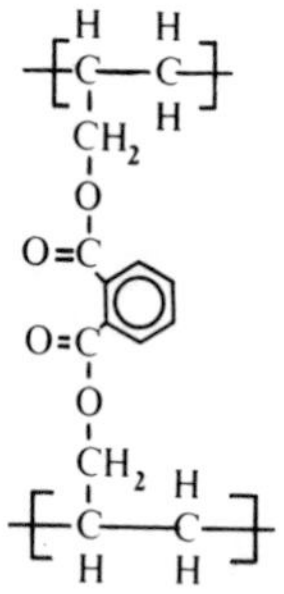

Network, addition

(monomer is 4-functional with two double bonds)

I

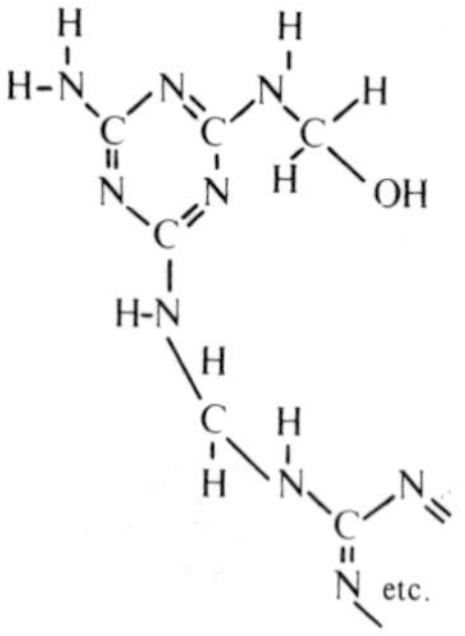

Network, condensation

Similar structures result when formaldehyde is condensed with urea or phenol

J $\left[\begin{array}{c} F\ F \\ C-C \\ F\ F \end{array}\right]_x$ Linear, addition

K $\left[O-\langle\rangle-\right]_x$ (with CH₃ groups) Linear, condensation "Oxidative coupling" H₂O out

L $\left[\begin{array}{c} H\ H \\ C-C \\ H\ CH_3 \end{array}\right]_x$ Linear, addition

M $\left[\begin{array}{c} H \\ C-O \\ H \end{array}\right]_x$ Linear, addition (Only other double bond that forms addition polymers)

N H$\left[O-\langle\rangle-\underset{CH_3}{\overset{CH_3}{C}}-\langle\rangle-O-\underset{O}{\overset{\ }{C}}\right]_x$Cl (HCl out) Linear, condensation

$\left[O-\overset{O}{\overset{\|}{C}}-O\right]$ Characteristic carbonate linkage

O H$\left[O-\langle\rangle-\underset{CH_3}{\overset{CH_3}{C}}-\langle\rangle-O-\overset{H\ H\ H}{\underset{H\ O\ H}{C-C-C}}\right]_x$ with -H

Note: Get both condensation -OH + Cl → HCl and epoxide ring scission

$$-OH + H_2C-\underset{H}{\overset{H}{C}}-\underset{H}{\overset{H}{C}}-Cl + HO- \longrightarrow -O-\overset{H\ O\ H}{\underset{H\ H\ H}{C-C-C}}-O- + HCl$$

Linear, condensation

for $x < 8$, liquid epoxy } Generally crosslinked later with diamines or
$8 < x < 20$, solid epoxy } acid anhydrides through -OH and terminal

$-\underset{O}{\overset{H}{C}}-CH_2$ groups

$x \approx 100$, linear, "phenoxy" plastic

P $\left[\begin{array}{c} CH_3 \\ Si-O \\ CH_3 \end{array}\right]_x$ (HCl out) Linear, condensation Commercial materials contain some

$Cl-\underset{Cl}{\overset{CH_3}{Si}}-Cl$ which is 3 functional and allows crosslinking Also acetate may be substituted for Cl

Characteristic urethane linkage

Q

R and R' vary widely—often R is already a low −x polymer

Linear, condensation nothing out, but diisocyanate can be considered

minus $2H_2O$

Can be crosslinked through amine groups, or by using higher functional alcohols or isocyanates.

R

Linear, condensation

Characteristic imide linkage (only example where both amine H's react)

S

(NaCl out)

CHAPTER 3

Bonding in Polymers

3.1 TYPES OF BONDS

Various types of bonding hold the atoms together in polymeric materials, as opposed to metals, for example, where only one type of bonding exists. These types are (1) primary covalent, (2) hydrogen bonding, (3) dipole interaction, (4) van der Waals, and (5) ionic, examples of which are shown in Fig. 3.1.[1] Hydrogen bonding, dipole interaction, van der Waals, and ionic bonding are known collectively as secondary forces. The distinctions are not always clear-cut; hydrogen bonds may be considered the extreme of dipole interactions, for example.

3.2 BOND DISTANCES AND STRENGTHS

Regardless of the type of bonding, the potential energy of the interacting atoms as a function of the separation between them is represented qualitatively by the potential function sketched in Fig. 3.2. As the interacting centers are brought together from large separation, an increasingly great attraction tends to draw them together (negative potential energy). Beyond the separation r_m, as the atoms are brought closer together, their electronic "atmospheres" begin to interact, and a powerful repulsion is set up. At r_m the system is at a potential energy minimum, that is, at its most probable or equilibrium separation, r_m being the equilibrium bond distance. The "depth" of the potential well, ϵ, is the energy required to break the bond, separating the atoms completely.

Table 3.1 lists the approximate bond strengths and interatomic distances of the bonds encountered in polymeric materials. The important fact to notice here is how much stronger the primary covalent bonds are than the others. As the material's temperature is raised and its thermal energy (kT) is thereby

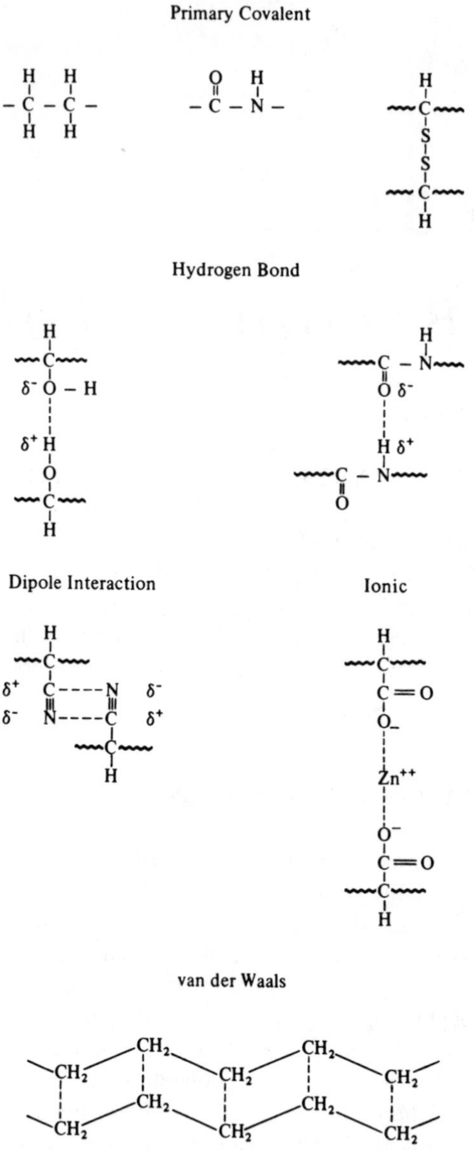

Primary Covalent

$$-\overset{\displaystyle H}{\underset{\displaystyle H}{C}}-\overset{\displaystyle H}{\underset{\displaystyle H}{C}}-$$

$$-\overset{\displaystyle O}{\overset{\|}{C}}-\overset{\displaystyle H}{\underset{\displaystyle}{N}}-$$

Hydrogen Bond

Dipole Interaction

Ionic

van der Waals

Figure 3.1   Bonding in Polymer Systems[1]

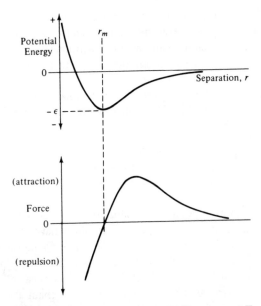

Figure 3.2   Interatomic Potential Energy and Force.

Table 3.1   Bond parameters[1,2]

Bond type	Interatomic distance r_m, Å	Dissociation energy, $-\epsilon$, kcal/mole
Primary covalent	1–2	50–200
Hydrogen bond	2–3	3–7
Dipole interaction	2–3	1.5–3
van der Waals	3–5	0.5–2
Ionic	2–3	10–20

increased, the primary covalent bonds will be the last to dissociate when the available thermal energy exceeds the dissociation energy.

3.3   BONDING AND RESPONSE TO TEMPERATURE

Now, in linear and branched polymers only the secondary bonds hold the individual polymer chains together (neglecting temporary mechanical entanglements). Thus, as the temperature is raised, a point will be reached where the forces holding the chains together become insignificant, and they are then free to slide past one another, that is, to *flow* upon the application of stress. Therefore *linear* and *branched* polymers are generally *thermoplastic*. The main chains in a crosslinked polymer, on the other hand, are held together by the same primary covalent bonds as the atoms in the main chains. When the thermal energy exceeds the dissociation energy of the primary covalent bonds, both main chain and crosslink bonds fail randomly, and the polymer degrades. Hence *crosslinked polymers are thermosetting.*

There are some exceptions to these generalizations. It is occasionally possible for secondary forces to make up for in quantity what they lack in quality. Polyacrylonitrile

$$\left(\begin{array}{c} H \ \ H \\ | \ \ \ | \\ -C-C- \\ | \ \ \ | \\ H \ \ C \\ \ \ \ \ \ ||| \\ \ \ \ \ \ N \end{array} \right)_x$$

, for example, is capable of strong interactions

at every other carbon atom along the chains. If these secondary forces could be broken one by one, that is, "unzipped," it would behave as a typical thermoplastic. This is impossible, of course, and by the time enough of the secondary bonds have been dissociated to free the chains and allow flow, the dissociation energy of some carbon–carbon main-chain bonds has been exceeded and the material degraded. Extreme stiffness of the polymer chain also contributes to this sort of behavior. Cellulose has a bulky, complex repeating unit that contains three hydroxyl groups. Though linear, its chains are therefore stiff and strongly hydrogen bonded, and it is not thermoplastic. If the hydroxyls are reacted with acids, for example, nitric, acetic, or butyric, the resulting cellulose esters are typical thermoplastics largely because of the reduced hydrogen bonding.

$$\left[R(OH)_3 \right]_x \ + \ 3x \ HO\text{-}\overset{\displaystyle O}{\overset{||}{O}}\text{-}CH_3 \ \rightarrow \ \left[R \ (O\text{-}\overset{\displaystyle O}{\overset{||}{C}}\text{-}CH_3)_3 \right]_x \ + \ 3x \ H_2O$$

Cellulose   Acetic acid   Cellulose acetate

Polytetrafluoroethylene (Teflon TFE) is another example due to the close packing and extensive secondary bonding of the main chains.

3.4   ACTION OF SOLVENTS

The action of solvents on polymers is in many ways similar to that of heat. Appropriate solvents, that is, those that can form secondary bonds to the polymer chains, can penetrate, replace the interchain secondary bonds, and thereby pull apart and dissolve linear and branched polymers. The polymer–solvent secondary bonds cannot overcome primary valence crosslinks, however; so crosslinked polymers are not soluble, although they may swell extensively. (Try soaking a rubber band in toluene overnight.) The amount of swelling is, in fact, a convenient measure of the extent of crosslinking. A lightly crosslinked polymer, a rubber band, for example, will swell tremendously, while one with extensive crosslinking, such as an ebonite ("hard rubber") bowling ball, will not swell noticeably at all.

REFERENCES

1. N. Platzer, *Ind. Eng. Chem.*, **61**, 5, 10 (1969).
2. M. L. Miller, *The Structure of Polymers*, Reinhold, New York, 1966.

CHAPTER 4 _____

Stereoisomerism

It is obvious now that the chemical nature of a polymer is of considerable importance in determining the polymer's properties. Of comparable significance is the way the molecules are arranged within the individual polymer chain.

4.1 INTRODUCTION

The carbon atom is normally (exclusively, for our purposes) tetravalent. In compounds such as methane (CH_4) and carbon tetrachloride (CCl_4), the four identical substituents surround it in a symmetrical tetrahedral geometry. If the substituent atoms are not identical, the symmetry is destroyed, but the general tetrahedral pattern is maintained. This is still true for each carbon atom in the interior of a linear polymer chain, where two of the substituents are chains. If a polyethylene chain were to be stretched out, for example, the carbon atoms in the chain backbone would lie in a zigzag fashion in a plane, with the hydrogen substituents on either side of the plane (Fig. 4.1). In the case of polyethylene with a perfectly symmetrical repeating unit, this structural arrangement is mostly of academic importance. With polymers in which the repeating unit is not symmetrical, however, it assumes great importance.

4.2 STEREOISOMERISM IN VINYL POLYMERS

Before beginning a discussion of structural arrangements in vinyl polymers, it must be pointed out that they polymerize almost exclusively in a head-to-tail configuration. The reasons for this are the steric interference of similar substituent groups and the electrostatic repulsion of groups with similar polarities. There are then three possible ways in which the unsymmetrical group may be

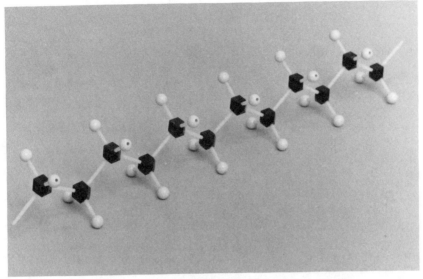

Figure 4.1   The Geometry of a Polyethylene Chain

arranged with respect to the carbon–carbon backbone plane (see Fig. 4.2), that is, three *stereoisomers*.

1. **Atactic.** A random arrangement of the unsymmetrical groups is known as an *atactic* structure. Its lack of regularity has important consequences.
2. **Isotactic.** The structure in which all of the groups are lined up on the same side of the backbone plane is termed *isotactic.*
3. **Syndiotactic.** Alternating placement of the group on either side of the chain is the *syndiotactic* structure.

The terms above were coined by Dr. Giulio Natta, who shared the 1964 Nobel Chemistry Prize for his work in the area.

Although useful for descriptive purposes, the planar zigzag arrangement of the main-chain carbon atoms is not always the one preferred by nature; that is, it is not necessarily the minimum free-energy configuration. In the case of isotactic and syndiotactic polypropylene, for example, the preferred (minimum-energy) configurations are quite regular, while the atactic is irregular (Fig. 4.3). Atactic polypropylene has a consistency somewhat like used chewing gum, whereas the stereoregular forms are hard, rigid plastics.

The type of stereoregularity described above is a direct result of the dissymmetry of vinyl monomers. It is established in the polymerization reaction, and no amount of twisting and turning the chain about its bonds can convert one stereoisomer into another (molecular models are a real help here).

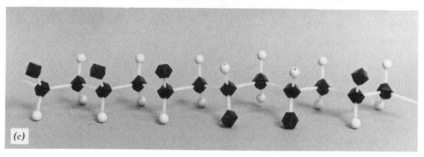

Figure 4.2   Stereoisomerism in Vinyl Polymers. a, isotactic; b, syndiotactic; c, atactic.

The situation is even more complex for monomers of the form HXC=CHX′. This is discussed in reference 1 but is currently of no commercial importance.

Example 1.   Both isotactic and atactic polymers of propylene oxide,

$$H-\underset{\diagdown}{\overset{\overset{\displaystyle H}{\vert}}{C}}-\underset{\diagup}{\overset{\overset{\displaystyle H}{\vert}}{C}}-CH_3$$

$$O$$

have been prepared by ring scission polymerization.

a.   Write the general structural formula for the polymer.

b.   Indicate how the atactic, isotactic, and syndiotactic structures differ.

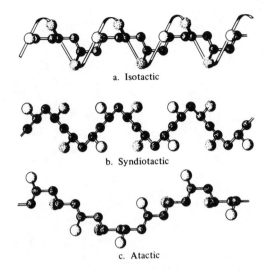

a. Isotactic

b. Syndiotactic

c. Atactic

Figure 4.3   Configurations of Polypropylene Chains. The large balls represent methyl ($-CH_3$) groups, and hydrogen atoms are not shown. From Ref. (1). Copyright © 1961 by Scientific American, Inc. All rights reserved.

Solution

A $\begin{array}{c} \text{H H} \\ | \quad | \\ -\!\!\left[\!\text{O}-\text{C}-\text{C}\!\right]_{x} \\ | \quad | \\ \text{H CH}_3 \end{array}$

B $-\text{O}-\overset{\text{H H}}{\underset{\text{H CH}_3}{\text{C}-\text{C}}}-\text{O}-\overset{\text{H H}}{\underset{\text{H CH}_3}{\text{C}-\text{C}}}-\text{O}-\overset{\text{H}\quad\text{CH}_3}{\underset{\text{H H}}{\text{C}-\text{C}}}-$ Atactic

$-\text{O}-\overset{\text{H H}}{\underset{\text{H CH}_3}{\text{C}-\text{C}}}-\text{O}-\overset{\text{H H}}{\underset{\text{H CH}_3}{\text{C}-\text{C}}}-\text{O}-\overset{\text{H H}}{\underset{\text{H CH}_3}{\text{C}-\text{C}}}-$ Isotactic

$-\text{O}-\overset{\text{H H}}{\underset{\text{H CH}_3}{\text{C}-\text{C}}}-\text{O}-\overset{\text{H}\quad\text{CH}_3}{\underset{\text{H H}}{\text{C}-\text{C}}}-\text{O}-\overset{\text{H H}}{\underset{\text{H CH}_3}{\text{C}-\text{C}}}-$ Syndiotactic

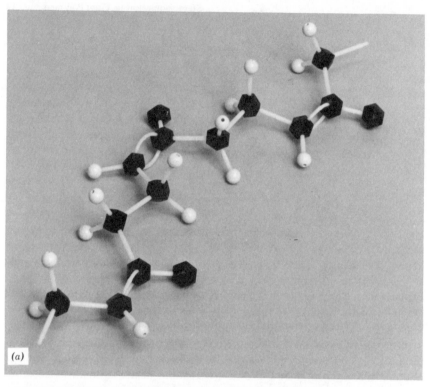

(a)

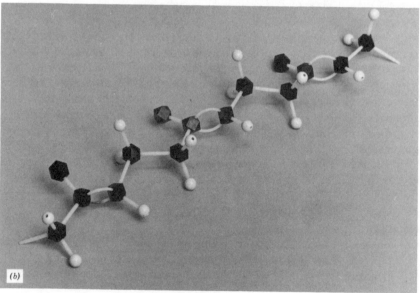

(b)

Figure 4.4 Stereoisomers of 1,4-polyisoprene. a, cis; b, trans.

4.3   STEREOISOMERISM IN DIENE POLYMERS

Another type of stereoisomerism arises in the case of poly 1,4-dienes because of the impossibility of rotation about a double bond. The substituent groups on the double-bonded carbons may either be on the same side of the chain (*cis*) or on opposite sides (*trans*), as shown in Fig. 4.4 for 1,4-polyisoprene. The chains of cis 1,4-polyisoprene assume a tortured, irregular configuration because of the steric interference of the substituents adjacent to the double bonds. This stereo-isomer is familiar as natural rubber. Trans 1,4-polyisoprene chains assume a regular structure. The polymer is gutta-percha, a hard, tough material, long used as a golf ball cover.

Note that stereoisomerism in poly 1,4-dienes does not depend on dissymmetry of the repeating unit. The same isomers are possible with butadiene, in which all carbon substituents are hydrogen.

Example 2.   Identify all the possible structural and stereoisomers that can result from the addition polymerization of butadiene,

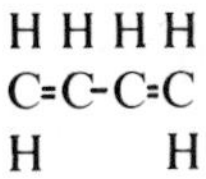

$$
\begin{array}{l}
\text{H H H H} \\
\text{C=C-C=C} \\
\text{H   H}
\end{array}
$$

Solution.   It will be recalled from chapter 2 that butadiene may undergo 1,4 addition or 1,2 addition. The 1,2 polymer can have cis and trans isomers like the

1,4 polyisoprene in Fig. 4.4. The 1,2 polymer, $\left[\begin{array}{cc} H & H \\ | & | \\ C & C \\ | & | \\ H & C-H \\ & \| \\ & H-C-H \end{array}\right]_x$ can have atactic,

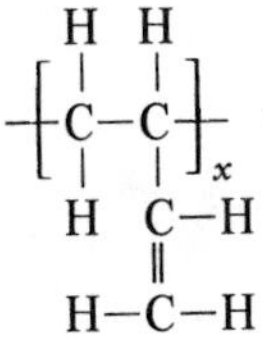

isotactic and syndiotactic stereoisomers like any vinyl polymer.

Example 3.   Identify all the possible structural and stereoisomers that can

result from the addition polymerization of chloroprene, $\overset{\displaystyle H}{\underset{\displaystyle H}{C}}=\overset{}{\underset{\displaystyle Cl}{C}}-\overset{\displaystyle H}{\underset{\displaystyle H}{C}}=\overset{\displaystyle H}{\underset{\displaystyle H}{C}}$ (the com-

mercial polymer of which is known as Neoprene).

Solution.   As in Example 2, the 1,4 polymer can have cis and trans stereo-isomers. This unsymmetrical diene monomer can also undergo both 1,2 addition

$$-\left[\begin{array}{cc} H & Cl \\ C - C \\ H & \underset{\|}{C}-H \\ & H-C-H \end{array}\right]_{x} \quad and\ 3,4\ addition \quad -\left[\begin{array}{cc} H & H \\ C - C \\ Cl-\underset{\|}{C} & H \\ H-C-H \end{array}\right]_{x}.$$ *Each* of these *structural* isomers may

have atactic, isotactic, or syndiotactic stereoisomers. Thus, in principle at least, there are *eight* different isomers of polychloroprene possible.

REFERENCES

1.   G. Natta, *Sci. Am.*, **205**, 2, 33 (1961).

CHAPTER 5

Crystallinity

It was pointed out in the previous chapter that the stereospecific arrangement of the atoms in polymer chains can exert a significant influence on the properties of the bulk polymer. In order that we may appreciate why this is so, the subject of polymer crystallinity is introduced here.

5.1 REQUIREMENTS FOR CRYSTALLINITY

Although the precise nature of crystallinity in polymers is still under investigation, a number of facts have long been known about the requirements for polymer crystallinity. First, an ordered, regular chain structure is necessary to allow the chains to pack into a regular crystal lattice. Thus stereoregular polymers are more likely to be crystalline than those that have irregular chain structures. Irregular, protruding side groups interfere with the arrangement of the main chains in a regular lattice and hinder crystallinity. Second, no matter how regular the chains, the secondary forces holding the chains together in the crystal lattice must be strong enough to overcome the disordering effect of thermal energy; so hydrogen bonding or strong dipole interactions promote crystallinity, and other things being equal, raise the crystalline melting point.

X-ray studies show that there are numerous polymers that do not meet the above criteria and show no traces of crystallinity; that is, they are completely *amorphous*. On the other hand, despite intensive efforts, no one has succeeded in producing a completely crystalline polymer. The crystalline content may in certain cases be pushed up to on the order of 98%, but there always remains a few percent amorphous material. In the case of metals, defect concentrations are on the order of parts per million; so they can be considered perfectly crystalline in comparison with polymers. The fiber industry takes advantage

37

of the fact that the degree of crystallinity of certain polymers can be increased by a "drawing" operation, that is, by stretching the fibers.

References 1 and 2 contain extensive discussions of the techniques used to study crystallinity in polymers, as well as reviews of recent research results.

5.2   THE FRINGED-MICELLE MODEL

The first attempt to explain the observed properties of crystalline (the word should be prefaced by "semi," but rarely is) polymers was the *fringed-micelle* model (Fig. 5.1). This model pictures crystalline regions known as fringed micelles or *crystallites* interspersed in an amorphous matrix. The crystallites, whose dimensions are on the order of several hundred angstroms, are small volumes in which portions of the chains are regularly aligned parallel to one another, tightly packed into a crystal lattice. The chains, however, are many times longer than the dimensions of an individual crystallite; so they pass from one crystallite through an amorphous area, back into another crystallite, and so on. This model explains nicely the coexistence of crystalline and amorphous material in polymers, and also explains the increase in crystallinity observed in the drawing operation. Stretching the polymer orients the chains in the direction of the stress, increasing the alignment in the amorphous areas and producing greater degrees of crystallinity (Fig. 5.1b). Since the chains pass at random from one crystallite to another, it is easily seen why perfect crystallinity can never be achieved. It also explains why the effects of crystallinity on

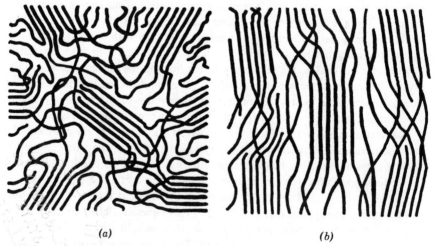

(a)   *(b)*

Figure 5.1   The Fringed Micelle Model. a, unoriented; b, chains oriented by applied stress.

mechanical properties are in many ways similar to those of crosslinking, because, like crosslinks, the crystallites tie the individual chains together. Unlike crosslinks, though, the crystallites will generally melt before the polymer degrades, and solvents that form extremely powerful secondary bonds with the chains can dissolve them.

The fringed-micelle model, although now largely superseded by more recent developments, still does a good job of predicting the effects of crystallinity on mechanical properties of polymers.

5.3   FOLDED-CHAIN CRYSTALLITES[1-4]

The first direct observations of the nature of polymer crystallinity resulted from the growth of single crystals from dilute solution. Either by cooling or evaporation of solvent, thin, pyramidal, or platelike polymer crystals (lamellae) were precipitated from dilute solutions (Fig. 5.2). These crystals were several hundred thousand angstroms along a side and about 100 Å thick. This was fine, except that x-ray measurements showed that the polymer chains were aligned perpendicular to the large flat faces of the crystals, and it was known that the individual chains were on the order of 1000 Å long (10^4 Å $= 1\ \mu$m). How could a chain fit into a crystal one-tenth its length? The only answer is that the chain must fold back on itself, as shown in Fig. 5.2.

This folded-chain model has been well substantiated for single polymer crystals. The lamellae are about 50 to 60 carbon atoms thick, with about five carbon atoms in a direct reentry fold. These atoms in the fold, of course, can never be part of a crystal lattice.

Similar structures exist in bulk polymer samples crystallized from the melt. Figure 5.3 illustrates a model combining the folded-chain lamellae with the interlamellar amorphous material tying the lamellae together in a bulk polymer.

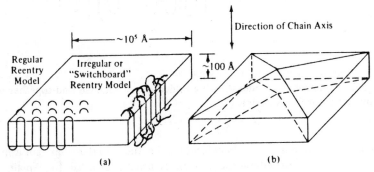

Figure 5.2   Polymer Single Crystals. a, flat lamellae; b, pyramidal lamellae. Two concepts of chain re-entry are illustrated.

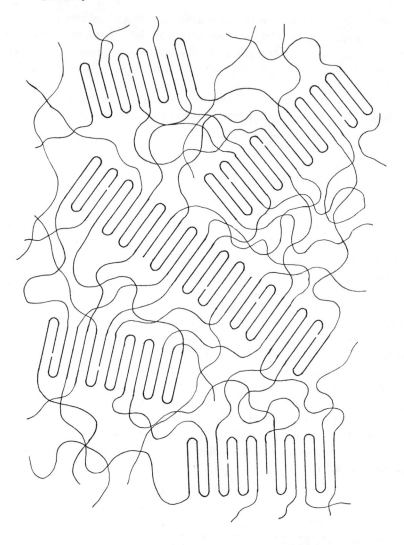

Figure 5.3 Compromise model showing folded-chain lamellae tied together by interlamellar amorphous chains.

Additional orientation and crystallization in the interlamellar amorphous regions, as in the fringed-micelle model, is usually invoked to explain the increase in crystallinity with drawing.

5.4   EXTENDED-CHAIN CRYSTALS

Recent work[5] has shown that polymers crystallized from a melt while subjected to extensional flow, which aligns the chains in the direction of extension, form fibrillar structures. These are believed to be extended-chain crystals, in which the chains are aligned parallel to one another over great distances, with a minimum of chain folding. Extended-chain crystals have also been identified as tying together folded-chain lamellae in bulk-crystallized polymers, and as forming the core of the interesting "shish kebab" structures grown from dilute solutions subjected to shearing.[1]

5.5   SPHERULITES

Not only are polymer chains often arranged to form crystallites, but these crystallites are often arranged in larger aggregates known as *spherulites*. These spherulites grow radially from a point of nucleation until other spherulites are encountered. Thus the size of the individual spherulites can be controlled by the number of nuclei present, more nuclei resulting in more but smaller spherulites. Spherulites are in some ways similar to the grain structure in metals. They are typically about 0.01 mm in diameter and have a "Maltese cross" appearance between crossed polaroids. Figure 5.4 shows how the polymer chains are thought to be arranged in the spherulites. Large spherulites contribute to brittleness in polymers. To avoid this, nucleating agents are often added or the polymer is shock cooled (which increases the nucleation rate) to promote smaller spherulites.

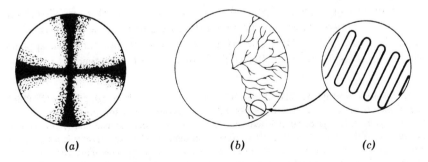

(a)   (b)   (c)

Figure 5.4   Spherulites. a, appearance between crossed polaroids; b, branching of lamellae; c, orientation of chains in lamellae.

5.6   THE EFFECTS OF CRYSTALLINITY ON POLYMER PROPERTIES

The presence of crystalline material in polymers strongly influences their properties. This influence is most profound on the mechanical properties.

Since the polymer chains are packed together more efficiently and tightly in the crystalline than in the amorphous areas, the crystallites will have a higher density. Thus "low-density" ($0.915 \, g/cm^3$) polyethylene is about 60% crystalline, while "high-density" ($0.97 \, g/cm^3$) polyethylene is about 95% crystalline. Density is, in fact, a convenient measure of the degree of crystallinity.

In the case of polyethylene, the differences in density arise from differences in the degree of branching. The branch points sterically prevent packing into a crystal lattice in their immediate vicinity and thus lower the degree of crystallinity.*

Since the polymer chains are more tightly packed in the crystalline than in the amorphous areas, there are more of them available per unit area to support a stress. Also, since they are in close and regular contact over relatively long distances in the crystallites, the net forces holding them together are far greater than in the amorphous regions. Thus crystallinity can significantly increase the strength and rigidity of a polymer. For this reason the stereoregular polypropylenes, which can and do crystallize, are hard and rigid plastics, while the irregular atactic polymer is amorphous and soft and sticky. The influence of the degree of crystallinity on some of the properties of polyethylene is illustrated in Table 5.1.

The mechanical properties in Table 5.1 apply basically to samples crystallized from a quiescent melt, which on a macroscopic level have a random overall orientation of chain and crystal axes. The materials mentioned in Section 5.4 show remarkable mechanical properties in the fibril direction, as the chains are most efficiently arrayed to support the stress. This approach offers great promise for obtaining "superstrength" polymers.

The optical properties of polymers also depend on crystallinity. When light passes between two phases with different refractive indices, some of it is scattered at the interface (e.g., a large salt crystal is transparent, but table salt appears white because light must pass alternately from air to salt many times). Crystalline

---

*Low-density polyethylene (LDPE) has traditionally been made by a high-pressure (25,000–50,000 psi) process and high-density polyethylene (HDPE) by a low-pressure (~ 100 psi) process. Thus LDPE is often referred to as high-pressure polyethylene and HDPE as low-pressure polyethylene. If that weren't confusing enough, low-density polyethylenes made by low-pressure processes have recently been commercialized. These new materials have most unfortunately (and inaccurately) been termed linear, low-density polyethylene (LLDPE). If they were truly linear, they wouldn't be low density. They are believed to have relatively short, straight branches arising from random copolymerization with α-olefins [$H_2C=CH(CH_2)_{\overline{n}}CH_3$] as compared to the long "branched branches" of traditional LDPE.

Table 5.1 The influence of crystallinity on some of the properties of polyethylene[a]

Commercial product	Low density	Medium density	High density
Density range, g/cm^3	0.910–0.925	0.926–0.940	0.941–0.965
Approximate % crystallinity	60–70	70–80	80–95
Branching, equivalent CH$_3$ groups/1000 carbon atoms	15–30	5–15	1–5
Crystalline melting point, °C	110–120	120–130	130–136
Hardness, Shore D	41–46	50–60	60–70
Tensile modulus, psi	0.14–0.38 × 10^5	0.25–0.55 × 10^5	0.6– 1.8 × 10^5
(N/m^2)	(0.97–2.6 × 10^8)	(1.7 –3.8 × 10^8)	(4.1–12.4 × 10^8)
Tensile strength, psi	600–2300	1200–3500	3100–5500
(N/m^2)	(0.41–1.6 × 10^7)	(0.83–2.4 × 10^7)	(2.1– 3.8 × 10^7)
Flexural modulus, psi	0.08–0.6 × 10^5	0.6 –1.15 × 10^5	1.0– 2.6 × 10^5
(N/m^2)	(0.34–4.1 × 10^8)	(4.1 –7.9 × 10^8)	(6.9–18 × 10^8)

[a] It must be kept in mind that mechanical properties are influenced by factors other than the degree of crystallinity (molecular weight, in particular).

polymers are actually two-phase systems, with a crystalline phase dispersed in an amorphous matrix. The denser crystalline areas have a higher refractive index than the amorphous areas; so crystalline polymers are either opaque or translucent. As the dimensions of the crystallites become appreciably smaller than the wavelength of light, the amount of light scattered decreases, and a polymer with very small crystallites and a low degree of crystallinity might appear transparent; but there is still some question as to whether such materials really exist. So, in general, *transparent polymers are completely amorphous.* The converse is not necessarily true; that is, lack of transparency in a polymer may be due to crystallinity, but it can also be caused by fillers. If the polymer is known to be a pure homopolymer, however, translucency is a sure sign of crystallinity. Thus commercial homopolystyrene, which is atactic and therefore completely amorphous because of the irregular arrangement of the bulky phenyl side groups, is perfectly transparent. Unfortunately, it is often called "crystal" polystyrene because of its "crystal clarity." Isotactic polystyrene has been synthesized in the laboratory and does crystallize. It has the white, translucent appearance typical of polyethylene. High-impact polystyrene consists of a dispersion of polybutadiene rubber particles in a continuous phase of atactic polystyrene. The two phases have different refractive indices; so the composite scatters light and also appears white and translucent, despite the fact that both phases are completely amorphous.

Example 1.   Explain the following facts:

a.   Polyethylene and polypropylene produced with stereospecific catalysts are each fairly rigid, translucent plastics, while a 65–35 copolymer of the two, produced in exactly the same manner, is a soft, transparent rubber.

b.   A plastic is similar in appearance and mechanical properties to the polyethylene and polypropylene described in (a), but it consists of 65% ethylene and 35% propylene units. The two components of this plastic cannot be separated by any physical or chemical means without degrading the polymer.

Solution

a.   The polyethylene produced with these catalysts is linear and thus highly crystalline. The polypropylene is isotactic and also highly crystalline. The crystallinity confers mechanical strength and translucency. The 65–35 copolymer, ethylene-propylene rubber (EPR), is a *random* copolymer; so the CH_3 groups from the propylene monomer are arranged at irregular intervals along the chain, preventing packing in a regular crystal lattice and giving an amorphous, rubbery polymer.

b.   Since the components cannot be separated, they must be chemically bound within the chains. The properties indicate a crystalline polymer; so the CH_3

groups from the propylene cannot be spaced irregularly, as in the random copolymer in (a). Thus these materials must be *block* copolymers of ethylene and stereoregular polypropylene, poly(ethylene-b-propylene). The long blocks of ethylene pack into a polyethylene lattice, and the propylene blocks pack into a polypropylene lattice.

Example 2.   Polyvinyl alcohol is made by the hydrolysis of polyvinyl acetate because the vinyl alcohol monomer is unstable.

Polyvinyl acetate   Polyvinyl alcohol   Acetic acid

$$\left[\begin{array}{c} H\ H \\ -C-C- \\ H\ O \\ | \\ C=O \\ | \\ CH_3 \end{array}\right]_x + H_2O \longrightarrow \left[\begin{array}{c} H\ H \\ -C-C- \\ H\ OH \end{array}\right]_x + H_3\ C\text{-}\overset{\overset{\textstyle O}{\|}}{C}\text{-}OH$$

The extent of reaction may be controlled to yield polymers with anywhere from 0 to 100% of the original acetate groups hydrolyzed. Pure polyvinyl acetate (0% hydrolyzed) is insoluble in water. It has been observed, however, that as the extent of hydrolysis is increased, the polymers become more water soluble up to about 87% hydrolysis, after which further hydrolysis ⌣creases water solubility at room temperature. Explain briefly.

Solution.   The normal polyvinyl acetate is atactic, and the irregular arrangement of the acetate side groups renders it completely amorphous. Water cannot form strong enough secondary bonds with the chains to dissolve it. As the acetate groups are replaced by hydroxyls, sites that can form strong hydrogen bonds with water are introduced, thereby increasing the solubility. At high degrees of substitution, replacement of the bulky acetate groups with the compact hydroxyls allows the chains to pack into a crystal lattice. The hydroxyl groups provide hydrogen bonding sites between the chains which help hold them in the lattice, and thus solubility is reduced.

Example 3.   Explain the following experiment: A weight is tied to the end of a polyvinyl alcohol fiber. The weight and part of the fiber are dunked in a beaker of boiling water. As long as the weight remains suspended, the situation is stable, but when the weight is rested on the bottom of the beaker, the fiber dissolves.

Solution.   As long as the weight is suspended from the fiber, the stress maintains the alignment of the chains in a crystal lattice, against the disordering effects of thermal energy and solvent (water) penetration. When the weight is rested on the bottom, the stress is removed, allowing the polymer to dissolve.

Example 4.   When a rubber band is rapidly stretched, if placed to the lips, it can be felt to have warmed. If it is held in the stretched configuration long enough to reach room temperature again and then suddenly is released, it will cool perceptibly. Explain.

Solution.   If the rubber band is stretched rapidly enough, the process will be adiabatic. Natural rubber, when stretched to high elongations, crystallizes as a result of chain alignment. As in other materials, crystallization is an exothermic process; so the energy given off warms the material. The reverse is observed when the band is released and the crystals melt.

REFERENCES

1.   J. M. Schultz, *Polymer Material Science,* Prentice-Hall, Englewood Cliffs, N.J., 1974.
2.   B. Wunderlich, *Macromolecular Physics,* Vol. 1: *Crystal Structure Morphology, Defects,* Academic Press, New York, 1973.
3.   P. H. Geil, *Polymer Single Crystals,* Interscience, New York, 1963.
4.   G. C. Oppenlander, *Science,* **159,** 1311 (1968).
5.   J. H. Southern and R. S. Porter, *J. Appl. Polym. Sci.,* **14,** 2305 (1970); *J. Macromol. Sci. – Phys.,* **B4,** 541 (1970).

CHAPTER **6** _____

Characterization of Molecular Weight

6.1 INTRODUCTION

With the exception of a few naturally occurring polymers, all polymers consist of molecules with a *distribution* of chain lengths. It is therefore necessary to characterize the entire distribution quantitatively, or at least to define and measure average chain lengths or molecular weights for these materials, as many important properties of the polymer depend on these quantities. An extensive review[1] that lists 304 references concerning the effects of molecular weight and molecular-weight distribution on the mechanical properties of polymers is available.

The concept of an *average* molecular weight causes some initial difficulty because we're used to thinking in terms of ordinary low-molecular-weight compounds in which the molecules are identical; there is a single, well-defined molecular weight for the compound. Where the molecules in a sample vary in size, however, the results depend on how you count.

In the case of pure, low-molecular-weight compounds, the molecular weight is defined as

$$M = \frac{W}{N} \tag{6.1}$$

where

W = total sample weight

N = number of moles in the sample

6.2 AVERAGE MOLECULAR WEIGHTS

Where a distribution of molecular weights exists, a *number-average* molecular weight $\bar{M}_n$ may be defined in a fashion analogous to (6.1):

47

$$\bar{M}_n = \frac{W}{N} = \frac{\sum\limits_{x=1}^{\infty} n_x M_x}{\sum\limits_{x=1}^{\infty} n_x} = \frac{n_1 M_1}{\Sigma n_x} + \frac{n_2 M_2}{\Sigma n_x} + \cdots = \sum_{x=1}^{\infty} \left(\frac{n_x}{N}\right) M_x \quad (6.2)$$

where

$$W = \text{total sample weight} = \sum_{x=1}^{\infty} w_x = \sum_{x=1}^{\infty} n_x M_x$$

w_x = total *weight* of x-mer

$$N = \text{total } number \text{ } of \text{ } moles \text{ in the sample (of all sizes)} = \sum_{x=1}^{\infty} n_x$$

n_x = *number of moles* of x-mer

M_x = molecular weight of x-mer

(n_x/N) = mole fraction of x-mer

Any analytical technique that determines the *number* of moles present in a sample of known weight, regardless of size, will give the number-average molecular weight.

Rather than count the number of molecules of each size present in a sample, it is possible to define an average in terms of the *weights* of molecules present at each size level. This is the weight-average molecular weight $\bar{M}_w$. A good way to illustrate the differences between the two averages is to consider the analogy of a mixture of variously sized ball bearings rolling down a trough into which successively larger slots have been cut (Fig. 6.1). The smallest ball bearings fall into a compartment beneath the first slot, the next larger size into a compartment beneath the second slot, and so on, with the last compartment holding the largest ball bearings. (We are assuming that there are a few discrete bearing sizes and, therefore, that each compartment holds bearings of a single diameter.) The subscript i serves to identify the compartments in order of increasing slot size. The number-average ball bearing diameter $\bar{D}_n$ is obtained by *counting* the numbers of ball bearings in each compartment,

$$\bar{D}_n = \frac{\Sigma n_i D_i}{\Sigma n_i}$$

where

D_i = diameter of bearings in compartment i

n_i = number of bearings in compartment i

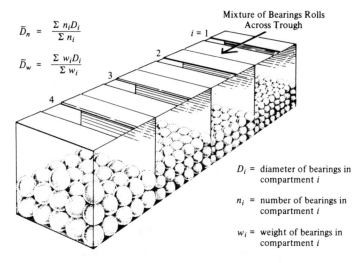

$$\bar{D}_n = \frac{\Sigma\, n_i D_i}{\Sigma\, n_i}$$

$$\bar{D}_w = \frac{\Sigma\, w_i D_i}{\Sigma\, w_i}$$

Mixture of Bearings Rolls
Across Trough

$i = 1$

D_i = diameter of bearings in compartment i

n_i = number of bearings in compartment i

w_i = weight of bearings in compartment i

Figure 6.1   Ball bearing analogy for average molecular weights.

and is analogous to the number-average molecular weight. An equally valid average diameter, the *weight-average* $\bar{D}_w$, is obtained by *weighing* the ball bearings in each compartment:

$$\bar{D}_w = \frac{\Sigma w_i D_i}{\Sigma w_i}$$

where

w_i = weight of all ball bearings in the ith compartment

D_i = diameter of the bearings in compartment i

The analogous *weight-average* molecular weight $\bar{M}_w$ is then

$$\bar{M}_w = \frac{\Sigma w_x M_x}{\Sigma w_x} = \frac{w_1 M_1}{\Sigma w_x} + \frac{w_2 M_2}{\Sigma w_x} + \cdots = \Sigma \left(\frac{w_x}{W}\right) M_x = \frac{\Sigma n_x M_x^2}{\Sigma n_x M_x}$$

(6.3)

where

w_x = weight of x-mer in sample = $n_x M_x$

w_x/W = weight fraction of x-mer in sample

Analytical procedures that, in effect, determine the weight of molecules at a given size level result in the weight-average molecular weight.

The number-average molecular weight is the *first moment* of the molecular-weight distribution, analogous to the center of gravity (the first moment of the mass distribution) in mechanics. The weight-average molecular weight, the *second moment* of the distribution, corresponds to the radius of gyration in mechanics. Higher moments, for example, $\bar{M}_z$, the third moment, may be defined and find occasional use.

It is sometimes more convenient to represent the dimensions of polymer molecules in terms of the degree of polymerization or chain length x, rather than molecular weight. They are simply related by

$$M_x = mx \tag{6.4a}$$

$$\bar{M}_n = m\bar{x}_n \tag{6.4b}$$

$$\bar{M}_w = m\bar{x}_w \tag{6.4c}$$

where

   m = molecular weight of a repeating unit

   $\bar{x}_n$ = number-average degree of polymerization or chain length

   $\bar{x}_w$ = weight-average degree of polymerization or chain length

(These relations neglect the end groups on the molecule. This is perfectly justifiable in most cases since the end groups are an insignificant portion of a typical large polymer molecule.) In terms of chain lengths, (6.2) and (6.3) are

$$\bar{x}_n = \frac{\Sigma n_x x}{\Sigma n_x} = \Sigma \left(\frac{n_x}{N}\right) x \tag{6.2a}$$

$$\bar{x}_w = \frac{\Sigma n_x x^2}{\Sigma n_x x} = \Sigma \left(\frac{w_x}{W}\right) x \tag{6.3a}$$

It may be shown that $\bar{M}_w \geqslant \bar{M}_n (\bar{x}_w \geqslant \bar{x}_n)$. The averages are equal only for a *monodisperse* (all molecules the same size) polymer. The ratio $\bar{M}_w/\bar{M}_n = \bar{x}_w/\bar{x}_n$ is known as the *polydispersity index* and is a measure of the breadth of the molecular-weight distribution. Values range from about 1.02 for carefully fractionated or anionic addition polymers to over 50 for some commercial polymers.

   Example 1.   Measurements on two essentially monodisperse fractions of a linear polymer, A and B, yield molecular weights of 100,000 and 400,000, respectively. Mixture 1 is prepared from one part by weight of A and two parts by weight of B. Mixture 2 contains two parts by weight of A and one of B. Determine the weight- and number-average molecular weights of mixtures 1 and 2.

Solution.   For mixture 1

$$n_A = \frac{1}{100,000} = 1 \times 10^{-5}$$

$$n_B = \frac{2}{400,000} = 0.5 \times 10^{-5}$$

$$\bar{M}_n = \frac{\Sigma n_i M_i}{\Sigma n_i} = \frac{(1 \times 10^{-5})(10^5) + (0.5 \times 10^{-5})(4 \times 10^5)}{1 \times 10^{-5} + 0.5 \times 10^{-5}} = 2.0 \times 10^5$$

$$\bar{M}_w = \Sigma \left(\frac{w_i}{W}\right) M_i = (\tfrac{1}{3})(1 \times 10^5) + (\tfrac{2}{3})(4 \times 10^5) = 3 \times 10^5$$

For mixture 2

$$n_A = \frac{2}{100,000} = 2 \times 10^{-5}$$

$$n_B = \frac{1}{400,000} = 0.25 \times 10^{-5}$$

$$\bar{M}_n = \frac{(2 \times 10^{-5})(10^5) + (0.25 \times 10^{-5})(4 \times 10^5)}{2 \times 10^{-5} + 0.25 \times 10^{-5}} = 1.33 \times 10^5$$

$$\bar{M}_w = (\tfrac{2}{3})(1 \times 10^5) + \tfrac{1}{3}(4 \times 10^5) = 2 \times 10^5$$

Example 2.   Two *polydisperse* samples are mixed in equal weights. Sample A has $\bar{M}_n = 100,000$ and $\bar{M}_w = 200,000$. Sample B has $\bar{M}_n = 200,000$ and $\bar{M}_w = 400,000$. What are $\bar{M}_n$ and $\bar{M}_w$ of the mixture?

Solution.   First, let us derive general expressions for calculating the averages of mixtures.

$$\bar{M}_n \equiv \frac{W}{N} = \frac{\sum_i W_i}{\sum_i N_i}$$

where the subscript i refers to the various polydisperse components of the mixture. Now, for a given component,

$$N_i = \frac{W_i}{\bar{M}_{ni}}$$

$$\bar{M}_n(\text{mixture}) = \frac{\sum\limits_{i} W_i}{\sum\limits_{i}\left(\dfrac{W_i}{\bar{M}_{ni}}\right)} \tag{6.5}$$

$$\bar{M}_w \cong \frac{\sum w_x M_x}{W} = \frac{\sum\limits_{i}\left(\sum\limits_{x} w_x M_x\right)_i}{\sum\limits_{i} W_i}$$

$$\bar{M}_{wi} = \frac{\left(\sum\limits_{x} w_x M_x\right)_i}{W_i}$$

$$\bar{M}_w(\text{mixture}) = \frac{\sum\limits_{i}(\bar{M}_{wi} W_i)}{\sum\limits_{i} W_i} = \sum\limits_{i}\left(\frac{W_i}{\sum\limits_{i} W_i}\right)\bar{M}_{wi} \tag{6.6}$$

where $(W_i/\sum\limits_{i} W_i)$ is the weight fraction of component i in the mixture.
In this case let $W_A = 1$ g and $W_B = 1$ g. Then

$$\bar{M}_n = \frac{W_A + W_B}{(W_A/\bar{M}_{n_A}) + (W_B/\bar{M}_{n_B})} = \frac{1 + 1}{(1/10^5) + (1/2 \times 10^5)} = 133{,}000$$

$$\bar{M}_w = \left(\frac{W_A}{W_A + W_B}\right)\bar{M}_{w_A} + \left(\frac{W_B}{W_A + W_B}\right)M_{w_B}$$

$$= (\tfrac{1}{2})2 \times 10^5 + (\tfrac{1}{2})4 \times 10^5 = 300{,}000$$

Note that even though the polydispersity index of each component of the mixture is 2.0, that of the mixture is greater, 2.25.

6.3   DETERMINATION OF AVERAGE MOLECULAR WEIGHTS

In this section we consider common techniques for measuring average molecular weights, with the object of gaining a qualitative understanding of their operation and an appreciation for their advantages and limitations. Additional theoretical and experimental details are available elsewhere.[2]

In general, techniques for the determination of average molecular weights fall into two categories: absolute and relative. In the former, measured quantities are theoretically related to the average molecular weight; in the latter, a quantity that is in some way related to molecular weight is measured, but the exact relation must be established by calibration with one of the absolute methods.

A.   Absolute Methods

End-Group Analysis

If the chemical nature of the end groups on the polymer chains is known, standard analytical procedures may sometimes be employed to determine the concentration of the end groups and thereby of the polymer molecules, giving directly the number-average molecular weight. For example, in a linear polyester formed from a stoichiometrically equivalent batch, there is, on the average, one unreacted acid group and one unreaction $-OH$ group per molecule. These groups may sometimes be analyzed by appropriate titration. If an addition polymer is known to terminate by disproportionation (see Chapter 10), there will be one double bond per every two molecules, which may be detectable quantitatively by halogenation or infrared measurements. Other possibilities include the use of a radioactively tagged initiator which remains in the chain ends.

These methods have one drawback in addition to the necessity of knowing the nature of the end groups. As the molecular weight increases, the concentration of end groups (number per unit volume) decreases, and the measurement sensitivity drops off rapidly. For this reason they are generally limited to the range $\bar{M}_n < 10,000$.

Colligative Property Measurements

When a solute is added to a solvent, it causes a change in the activity and chemical potential (partial molal Gibbs free energy) of the solvent. The magnitude of the change is directly related to the solute concentration. For example, when pure water is boiled, the chemical potentials of the liquid and the vapor in equilibrium with it are the same. If, now, some salt is added to the water, it lowers the chemical potential of the liquid water. In order to reestablish equilibrium with the pure water vapor above the salt solution, the temperature of the system must be raised, causing a boiling-point elevation. In a similar fashion the addition of ethylene glycol antifreeze depresses the freezing point of water.

Freezing-point depression, boiling-point elevation, and a third technique, osmotic pressure, may be used to determine the number of moles of polymer per unit volume of solution and thereby establish the number-average molecular weight. The relevant thermodynamic equations for the three techniques are

$$\lim_{c \to 0} \frac{\Delta T_b}{c} = \frac{RT^2}{\rho \Delta H_v \bar{M}_n} \qquad \text{(boiling-point elevation)} \qquad (6.7)$$

$$\lim_{c \to 0} \frac{\Delta T_f}{c} = \frac{-RT^2}{\rho \Delta H_f \bar{M}_n} \qquad \text{(freezing-point depression)} \qquad (6.8)$$

$$\lim_{c \to 0} \frac{\pi}{c} = \frac{RT}{\bar{M}_n} \qquad \text{(osmotic pressure)} \qquad (6.9)$$

where

c = solute (polymer) concentration, mass/volume

T = absolute temperature

R = gas constant

ΔH_v = solvent enthalpy of vaporization

ΔH_f = solvent enthalpy of fusion

ΔT_b = boiling-point elevation

ΔT_f = freezing-point depression

π = osmotic pressure

ρ = density

It is important to note that the thermodynamic equations apply only for ideal solutions, a condition that can only be reached in the limit of infinite dilution of the solute.

Freezing-point depression and boiling-point elevation require precise measurements of very small temperature differences. Although they are used occasionally, the difficulties involved have prevented their widespread application. Osmotic pressure, on the other hand, is the most common method of determining $\bar{M}_n$. A schematic diagram of an osmometer is shown in Fig. 6.2. The solution and solvent chambers are separated by a "semipermeable" membrane, one that ideally allows passage of solvent molecules but not solute molecules. The solvent flows through the membrane to dilute the solution. This is a natural consequence of the tendency of the system to increase its entropy, which is accomplished by the dilution of the solution. This dilution continues until the tendency toward further dilution is counterbalanced by the increased pressure in the solution chamber. At this point, the chemical potential of the solvent is the same in both chambers, and the pressure difference between the chambers is the osmotic pressure π. By making measurements at several concentrations,

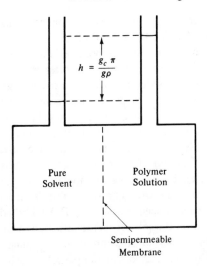

$$h = \frac{g_c \, \pi}{g\rho}$$

Pure
Solvent

Polymer
Solution

Semipermeable
Membrane

Figure 6.2   Schematic diagram of an osmometer.

plotting π/c versus c, and extrapolating to zero concentration, the number-average molecular weight is established through (6.9).

High-speed, automated membrane osmometers monitor the fluid volume in one of the chambers and externally *apply* the osmotic pressure to the solution chamber to *prevent* flow. Since no flow through the membrane is necessary, they can reduce measurement time from days to hours or even minutes.

One of the major difficulties with membrane osmometry is finding suitable semipermeable membranes. Ordinary cellophane or modifications of it are commonly used. Unfortunately, if the membrane is sufficiently "tight" to prevent the passage of low-molecular-weight chains, the rate of solvent passage is slow, and it takes longer to reach equilibrium. In practice, all membranes allow some of the low-molecular-weight polymer in a distribution to "sneak through." Also, as the average molecular weight increases, π decreases; so the measurement precision decreases. These factors usually limit the applicability of the technique to $50,000 < \bar{M}_n < 1,000,000$.

A related colligative technique is vapor-pressure osmometry. Two thermistors are placed in a carefully thermostatted chamber with pure solvent at the bottom so that the atmosphere is saturated with solvent vapor. A drop of solvent is placed on one thermistor, and a drop of polymer solution on the other. Because of the solvent's lower chemical potential in the solution, solvent vapor condenses on the solution drop, giving up its heat of condensation and warming the solution drop relative to the pure solvent drop. In principle, the equilibrium ΔT is thermodynamically related to the molar solution concentration, thereby allowing calculation of $\bar{M}_n$. In practice, heat losses (mainly along the thermistor leads) require that the instrument be calibrated for precise results, really making it a relative technique. On a routine basis, commerical instruments are probably limited to maximum $\bar{M}_n$'s of 40,000–50,000.[3]

The techniques discussed to this point establish the *number* of molecules present per unit volume of solution, regardless of their size or shape. Other methods measure quantities that are related to the average *mass* of the molecules in solution, thereby giving the weight-average molecular weight. One of the more common of these is *light scattering*, which is based on the fact that the intensity of light scattered by a polymer molecule is proportional, among other things, to the square of its mass. A light-scattering photometer measures the intensity of scattered light as a function of the scattering angle. Measurements are made at several concentrations. By a double extrapolation to zero angle and zero concentration (*Zimm plot*), and with a knowledge of the dependence of the solution refractive index versus concentration, $\bar{M}_w$ is established. This technique also provides information on the solvent–solute interaction and on the configuration of the polymer molecules in solution, since the quantitative nature of the scattering depends also on the size of the particles. Light scattering is generally applicable over the range $10,000 < \bar{M}_w < 10,000,000$.

The weight-average molecular weight can also be obtained with an ultracentrifuge, which distributes the molecules according to their mass in a centrifugal force field.

B.   Relative methods

The molecular-weight determination techniques discussed to this point allow direct calculation of the average molecular weight from experimentally measured quantities through known theoretical relations. Sometimes, however, these relations are not known, and although something that is known to be related to molecular weight is measured, one of the absolute methods above must be used to calibrate the technique.

A case in point is solution viscosity. It has long been known that relatively small amounts of dissolved polymer could cause tremendous increases in viscosity, and it is logical to assume that at a given concentration the larger molecules will impede flow more and give the higher viscosity. A quantitative basis for the treatment of solution viscosity was provided by Einstein in 1906.[4] For a suspension of *rigid, noninteracting spheres*, he developed the relation

$$\eta_r = \frac{\eta}{\eta_s} = 1 + 2.5\phi \tag{6.10}$$

where

η = viscosity of the suspension

η_s = viscosity of the solvent

η_r = *relative viscosity*

ϕ = volume fraction of spheres

Rearranging,

$$\left(\frac{\eta}{\eta_s} - 1\right) = \eta_{sp} = 2.5\phi \qquad (6.11)$$

where

$$\eta_{sp} = \text{specific viscosity}$$

Einstein's equation has been amply verified for very dilute suspensions of rigid spheres, where the assumptions in the derivation are met.

Polymer molecules in solution, of course, are not rigid, nor are they necessarily spherical, and at finite concentrations they are bound to interact with one another. All these factors depend on the interaction between the solute and solvent molecules, and hence the particular polymer–solvent system and the temperature. When these limitations are recognized, (6.11) may be rewritten

$$\left(\frac{\eta}{\eta_s} - 1\right) = \eta_{sp} = Kc \qquad (6.12)$$

where K is a function of the size of the dissolved molecules, their shape, their rigidity, the interactions between them, and the proportionality between volume fraction and concentration. Ultimately, then, K depends on the particular polymer–solvent system, the temperature, and the size of the molecules in solution. Dividing both sides by c

$$\left(\frac{(\eta/\eta_s) - 1}{c}\right) = \eta_{red} = K \qquad (6.13)$$

where

$$\eta_{red} = \text{reduced viscosity}$$

The effects of polymer–polymer interactions on the reduced viscosity may be eliminated by extrapolation to zero concentration:

$$\lim_{c \to 0}\left(\frac{\eta_{sp}}{c}\right) = [\eta] \qquad (6.14)$$

where $[\eta] = $ intrinsic viscosity.

The intrinsic viscosity, then, should be a function of the size of the polymer molecules in solution, the polymer–solvent system, and the temperature, and if measurements are made at constant temperature using a specified solvent for a particular polymer, it should be related to the polymer's molecular weight.

Huggins proposed a relation between reduced viscosity and concentration for dilute polymer solutions ($\eta_r < 2$):

$$\frac{\eta_{sp}}{c} = [\eta] + k'[\eta]^2 c \qquad \text{(Huggins equation)} \qquad (6.15a)$$

Interestingly enough, k' turns out to be approximately equal to 0.4 for a variety of polymer–solvent systems, providing a convenient means of estimating dilute solution viscosity versus concentration if the intrinsic viscosity is known. By expanding the natural logarithm in a power series, it may be shown that an equivalent form of the Huggins equation is

$$\eta_{inh} = \frac{\ln \eta_r}{c} = [\eta] + k''[\eta]^2 c \tag{6.15b}$$

where

$$\eta_{inh} = \text{inherent viscosity}$$

$$k'' = k' - 0.5$$

Plots of the reduced and inherent viscosities are linear with concentration, at least below concentrations of about 0.5 g/dl, as predicted by the Huggins equation, with a common intercept, the intrinsic viscosity (Fig. 6.3). Exceptions occur with polyelectrolytes, where the degree of ionization and therefore the chemical nature of the polymer changes with concentration.

Note that the intrinsic viscosity has dimensions of reciprocal concentration. For some strange reason, concentrations are usually given in grams per deciliter (= 100 ml). In fact, the relative, specific, reduced, intrinsic, and inherent viscosities are not true viscosities and do not have dimensions of viscosity. More appropriate terminology has been proposed, but has not been widely adopted. Table 6.1 summarizes the various quantities defined and typical units.

Now that the intrinsic viscosity has been established, how is it related to molecular weight? Studies of the intrinsic viscosity of essentially monodisperse polymer fractions whose molecular weights have been established by one of the absolute methods indicate a rather simple relation (Fig. 6.4):

$$[\eta]_x = K(M_x)^a \qquad (0.5 < a < 1.0) \tag{6.16}$$

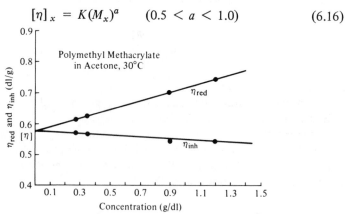

Figure 6.3   Determination of intrinsic viscosity (Example 3).

Table 6.1 Solution viscosity terminology[5]

Quantity	Common units	Common name	Recommended name
η	Centipoise	Solution viscosity	Solution viscosity
η_s	Centipoise	Solvent viscosity	Solvent viscosity
$\eta_r = \eta/\eta_s$	Dimensionless	Relative viscosity	Viscosity ratio
$\eta_{sp} = (\eta - \eta_s)/\eta_s = \eta_r - 1$	Dimensionless	Specific viscosity	—
$\eta_{red} = \eta_{sp}/c = \eta_r - 1/c$	Deciliters/gram	Reduced viscosity	Viscosity number
$\eta_{inh} = \ln \eta_r/c$	Deciliters/gram	Inherent viscosity	Logarithmic viscosity number
$[\eta] = \lim_{c \to 0} \eta_{red} = \lim_{c \to 0} \eta_{inh}$	Deciliters/gram	Intrinsic viscosity	Limiting viscosity number

59

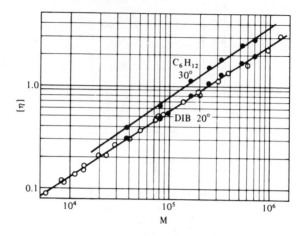

Figure 6.4 Intrinsic viscosity–molecular weight relations for polyisobutylene in cyclohexane at $30°C$ and diisobutylene at $20°C$. Reprinted from Paul J. Flory: PRINCIPLES OF POLYMER CHEMISTRY. Copyright 1953 by Cornell University. Used by permission of Cornell University Press.

where the subscript x refers to a monodisperse sample of a particular molecular weight. What about an unfractionated polydisperse sample? Experimentally, the measured intrinsic viscosity of a mixture of monodisperse fractions is a weight average;

$$[\eta] = \frac{\Sigma [\eta]_x w_x}{\Sigma w_x} = \Sigma \left(\frac{w_x}{W} \right) [\eta]_x \qquad (6.17)$$

A *viscosity-average* molecular weight $\bar{M}_v$ is defined in terms of this measured intrinsic viscosity:

$$\bar{M}_v = \left(\frac{[\eta]}{K} \right)^{1/a} = \left\{ \frac{\Sigma M_x^a w_x}{W} \right\}^{1/a} = \left\{ \frac{\Sigma M_x^a n_x M_x}{\Sigma n_x M_x} \right\}^{1/a} = \left\{ \frac{\Sigma n_x M_x^{(1+a)}}{\Sigma n_x M_x} \right\}^{1/a}$$

$$(6.18)$$

With $0.5 < a < 1.0$, $\bar{M}_n < \bar{M}_v < \bar{M}_w$, but $\bar{M}_v$ is closer to $\bar{M}_w$ than $\bar{M}_n$. If the molecular-weight distribution of a series of samples does not differ too much, that is, the ratios of the various averages remain the same, approximate equations of the type $[\eta] = K'(\bar{M}_w)^{a'}$ may be applicable.

Example 3.   The following data were obtained for a sample of polymethyl methacrylate in acetone at $30°C$:

η_r	c, g/100 ml
1.170	0.275
1.215	0.344
1.629	0.896
1.892	1.199

For polymethyl methacrylate in acetone at $30°C$, $[\eta] = 5.83 \times 10^{-5}\ (\bar{M}_v)^{0.72}$. Determine $[\eta]$ and $\bar{M}_v$ for the sample and k', the constant in the Huggins equation.

Solution.　For the data above, calculations give

η_{sp}	η_{red} (dl/g)	$\ln \eta_r$	η_{inh} (dl/g)
0.170	0.618	0.157	0.571
0.215	0.625	0.195	0.567
0.629	0.702	0.488	0.545
0.892	0.744	0.638	0.532

In Fig. 6.3, η_{red} and η_{inh} are plotted against concentration and extrapolated to a common intercept at zero concentration, the intrinsic viscosity, $[\eta] = 0.577$ dl/g:

$$\bar{M}_v = \left(\frac{[\eta]}{5.83 \times 10^{-5}} \right)^{1/0.72} = \left(\frac{0.577}{5.83 \times 10^{-5}} \right)^{1.39} = 355,000$$

Using the third point in the Huggins equation (6.15a), 0.702 dl/g = 0.577 dl/g + $k'(0.577$ dl/g$)^2(0.896$ g/dl$)$. Solving for k' gives $k' = 0.42$ (dimensionless).

Example 4.　By assuming that the fractions in Example 1 are polymethyl methacrylate, calculate $\bar{M}_v$ for mixtures 1 and 2 in acetone at $30°C$ and compare with $\bar{M}_n$ and $\bar{M}_w$ (a is given in Example 3).

Solution.　From (6.18), for mixture 1

$$\bar{M}_v = \left\{ \Sigma \left(\frac{w_x}{W} \right) M_x^a \right\}^{1/a} = \{(\tfrac{1}{3})(1 \times 10^5)^{0.72}$$

$$+ (\tfrac{2}{3})(4 \times 10^5)^{0.72}\}^{1/0.72} = 288,000$$

$$\bar{M}_n = 200,000$$
$$\bar{M}_w = 300,000$$

for mixture 2

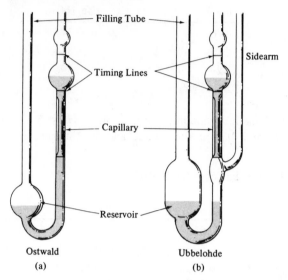

Figure 6.5   Dilute-solution viscometers. (a) Ostwald; (b) Ubbelohde.

$$\bar{M}_v = \{(\tfrac{2}{3})(1 \times 10^5)^{0.72}$$

$$+ (\tfrac{1}{3})(4 \times 10^5)^{0.72}\}^{1/0.72} = 187,000$$

$$\bar{M}_n = 133,000$$

$$\bar{M}_w = 200,000$$

Viscosities[*] for molecular-weight determination usually are measured in glass capillary viscometers, in which the solution flows through a capillary under its own head. Two common types, the Ostwald and Ubbelohde, are sketched in Fig. 6.5. Flow times t are related to the viscosity of the solution by an equation of the form

$$\frac{\eta}{\rho} = \nu = at + \frac{b}{t} \tag{6.19}$$

where a and b are instrument constants, ρ the solution density, and ν the kinematic viscosity. The last term, the kinetic energy correction, is generally negligible for flow times of over a minute, and since the densities of the dilute polymer solutions differ little from that of the solvent,

$$\frac{\eta}{\eta_s} \simeq \frac{t}{t_s} \tag{6.20}$$

[*]Since polymer solutions are non-Newtonian, intrinsic viscosity must be defined, strictly speaking, in terms of the zero-shear or lower Newtonian viscosity (Chapter 15). This is rarely a problem because the low shear rates in the usual glassware viscometers give just that. Occasionally, however, extrapolation to zero-shear conditions is required.

The Ubbelohde viscometer has the distinct advantage that the driving fluid head is independent of the amount of solution in it; hence dilution of the solutions can be carried out right in the instrument.

The equipment necessary for intrinsic viscosity determination is inexpensive, and the measurements straightforward and rapid. The Mark–Houwink constants K and a in (6.16) and (6.18) are extensively tabulated for a wide variety of polymer–solvent systems and temperatures.[6]

6.4   MOLECULAR-WEIGHT DISTRIBUTIONS

A typical synthetic polymer might consist of a mixture of molecules with degrees of polymerization x ranging from one to perhaps many million. The complete molecular-weight distribution specifies the mole (number) or mass (weight) fraction of molecules at each size level in a sample. (Actually, moles or masses could be specified, but they vary linearly with sample size; i.e., they are extensive quantities, while the fractions are intensive, independent of sample size, and therefore preferable.)

Distributions are often presented in the form of a plot of mole (n_x/N) or mass (w_x/W) fraction of x-mer versus *either x or M_x*. Since x and M_x differ by a constant factor, the molecular weight of the repeating unit m (6.4a), it makes little difference which is used. Because x can assume only integral values, a true distribution must consist of a series of spikes, one at each integral value of x, or separated by m molecular weight units if plotted against M_x. The height of the spike represents the mole or mass fraction of that particular x-mer. No analytical technique is capable of resolving the individual x-mers; so distributions are drawn (and represented mathematically) as continuous curves, the locus of the spike tops. This is sketched in Fig. 6.6.

The averages may be calculated from either the mole- or mass-fraction distributions. For continuous distributions, the summations in (6.2) and (6.3) must be replaced by the corresponding integrals (the subscript x has been dropped for clarity and because of the continuous nature of the distributions).

From the mole- (number-) fraction distribution (n/N)

$$\bar{M}_n = \text{first moment} = \frac{\int_0^\infty nM\,dM}{\int_0^\infty n\,dM} = \frac{\int_0^\infty (n/N)M\,dM}{\int_0^\infty (n/N)\,dM} \tag{6.21}$$

since $\bar{M}_n = m\bar{x}_n$ and $M = mx$

$$\bar{x}_n = \int_0^\infty \left(\frac{n}{N}\right)x\,dx \tag{6.22}$$

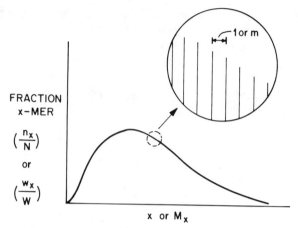

Figure 6.6   Molecular-weight distributions illustrating the actual discrete distribution and the usual continuous approximation to it.

Note that

$$\int_0^\infty \left(\frac{n}{N}\right) dx = 1 \tag{6.23}$$

that is, the mole fractions must sum to 1.

$$\bar{M}_w = \text{second moment} = \frac{\int_0^\infty nM^2\, dM}{\int_0^\infty nM\, dM} = \frac{\int_0^\infty (n/N)M^2\, dM}{\int_0^\infty (n/N)M\, dM} \tag{6.24}$$

since $\bar{M}_w = m\bar{x}_w$

$$\bar{x}_w = \frac{\int_0^\infty (n/N)x^2\, dx}{\int_0^\infty (n/N)x\, dx} = \frac{1}{\bar{x}_n} \int_0^\infty \left(\frac{n}{N}\right)x^2\, dx \tag{6.25}$$

From the mass- (weight-) fraction distribution (w/W)

$$\bar{M}_n = \frac{\int_0^\infty nM\, dM}{\int_0^\infty n\, dM} = \frac{\int_0^\infty w\, dM}{\int_0^\infty (w/M)\, dM} = \frac{\int_0^\infty (w/W)\, dM}{\int_0^\infty (1/M)(w/W)\, dM} \tag{6.26}$$

$$\bar{x}_n = \frac{\int_0^\infty (w/W)\, dx}{\int_0^\infty (1/x)(w/W)\, dx} \tag{6.27}$$

$$\bar{M}_w = \frac{\int_0^\infty nM^2\, dM}{\int_0^\infty nM\, dM} = \frac{\int_0^\infty (nM)M\, dM}{\int_0^\infty (nM)\, dM} = \frac{\int_0^\infty wM\, dM}{\int_0^\infty w\, dM} = \frac{\int_0^\infty (w/W)M\, dM}{\int_0^\infty (w/W)\, dM} \tag{6.28}$$

$$\bar{x}_w = \int_0^\infty \left(\frac{w}{W}\right) x \, dx \tag{6.29}$$

Note that

$$\int_0^\infty \left(\frac{w}{W}\right) dx = 1 \tag{6.30}$$

that is, the mass fractions must sum to 1.

Given the mole- (number-) fraction distribution, the mass- (weight) fraction distribution may be calculated, and vice versa. Since $w = nM$ and, by definition of $\bar{M}_n$ (6.2), $W = N\bar{M}_n$,

$$\left(\frac{w}{W}\right) = \frac{M}{\bar{M}_n}\left(\frac{n}{N}\right) = \frac{x}{\bar{x}_n}\left(\frac{n}{N}\right) \tag{6.31}$$

Example 5.   One analytic representation of a distribution of chain lengths which finds some practical application (as we will see later) is

$$\left(\frac{n}{N}\right) = \frac{1}{C} e^{-x/C}$$

where C is a constant.

a.   Determine $\bar{x}_n$ for this distribution.
b.   Determine $\bar{x}_w$ and the polydispersity index $\bar{x}_w/\bar{x}_n$ for this distribution.
c.   Obtain the expression for the weight-fraction distribution (w/W).
d.   Sketch the number- and weight-fraction distributions in the form $(n/N)\bar{x}_n$ and $(w/W)\bar{x}_n$ versus $x/\bar{x}_n$.

Solution.

a. $\bar{x}_n = \int_0^\infty (n/N) x \, dx = 1/C \int_0^\infty x e^{-x/C} \, dx$

Fortunately for most of us, the above definite integral is tabulated in standard references. The result is

$$\bar{x}_n = C$$

b. $\bar{x}_w = \dfrac{n}{\bar{x}_n} \displaystyle\int_0^\infty \left(\dfrac{n}{N}\right) x^2\, dx = \dfrac{1}{\bar{x}_n^2} \displaystyle\int_0^\infty x^2 e^{-x/\bar{x}_n}\, dx$

This integral is also tabulated and gives the simple result

$$\bar{x}_w = 2\bar{x}_n$$

that is, $\bar{x}_w/\bar{x}_n = 2$ for this distribution.

c. $\left(\dfrac{w}{W}\right) = \left(\dfrac{n}{N}\right)\dfrac{x}{\bar{x}_n} = \dfrac{x}{\bar{x}_n^2} e^{-x/\bar{x}_n}$

d.   The distributions are shown in Fig. 6.7. Note that while the *number* of molecules decreases exponentially with x, the *weight* of a molecule increases

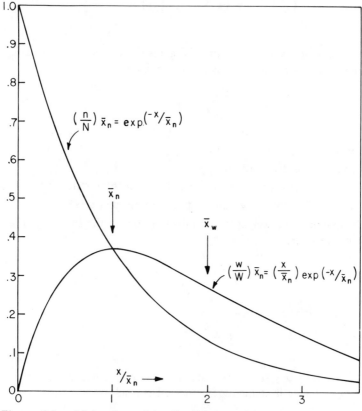

Figure 6.7   Molecular-weight distribution of Example 5, plotted in reduced form.

linearly with x. The weight fraction therefore goes through a maximum in this case. This distribution is derived for a particular kind of polymerization reaction in Section 10.7.

6.5   GEL-PERMEATION CHROMATOGRAPHY (GPC)*

Until fairly recently the approximate molecular-weight distribution of a polymer could only be determined by laborious fractionation of the sample according to molecular weight, followed by determination of the molecular weights of the individual fractions with one of the techniques previously discussed. The bases for these fractionation techniques are discussed in Chapter 7. Suffice it to say that they tend to be difficult and time consuming, and so, in general, are avoided whenever possible.

In the past decade or so, gel permeation chromatography has been firmly established as a means of rapidly determining molecular-weight averages and distributions. GPC makes use of a column, or series of columns, packed with particles of a porous substrate. The term gel refers to a crosslinked polymer that is swollen by the solvent used. This is perhaps the most common type of substrate, but others, for example, porous glass beads, are used. The column is maintained at a constant temperature, and solvent is passed through it at a constant rate. At the start of a run, a small amount of polymer solution is injected just ahead of the column. The solvent flow carries the polymer through the column. The smaller molecules in the sample have easy access to the substrate pores and diffuse in and out of the pores, following a circuitous route as they progress through the column. The large molecules simply can't fit into the pores and are swept more or less directly through the interstices in the packing. Thus a separation is obtained, the largest molecules being washed through the column first, followed by successively smaller ones.

A concentration-sensitive detector is placed at the outlet of the column. The most common detector is a differential refractometer, which measures the difference in refractive index between pure solvent and the polymer solution leaving the column, a sensitive measure of polymer concentration. Ultraviolet or infrared detectors can also be used if the polymer has some group that absorbs the radiation. Regardless of the type of detector, it is essential that it measure some quantity Q that is proportional only to the *mass* concentration of polymer at the column outlet (Q *must* be independent of M).

$$Q = kc \qquad (6.32)$$

---

*The same basic technique is also referred to as LEC (liquid exclusion chromatography), SEC (size exclusion chromatography), HPSEC (high-performance size exclusion chromatography), and so on.

where

$$Q = \text{detector readout}$$

$$k = \text{proportionality constant}$$

$$c = \textit{mass} \text{ concentration of polymer, g/cm}^3$$

Thus a GPC curve consists of a plot of Q (usually in arbitrary scale divisions) versus v, the volume of solvent that has passed through the detector since sample injection, called elution volume. Generally, the chart is "blipped" every 5 cm³ of elution volume.

A GPC curve is shown in Fig. 6.8. Such a curve is often immediately useful for quality control purposes. For example, if a curve is available for a material that performs acceptably in a particular application, the GPC curves for subsequent batches may be checked against it, revealing qualitative variations in molecular-weight distribution from the standard. This may be enough to take appropriate corrective action.

GPC is also useful in detecting low-molecular-weight additives, such as plasticizers or stabilizers. These materials show up as peaks at the low-M (high-v) end of the spectrum.

GPC is a relative method. To provide quantitative results, the relation between

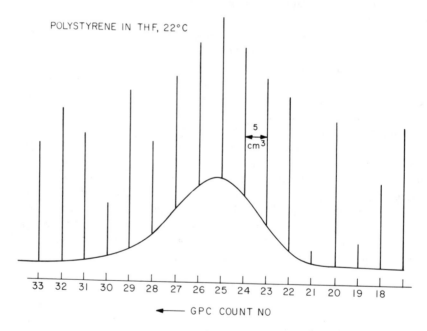

POLYSTYRENE IN THF, 22°C

5 cm³

33 32 31 30 29 28 27 26 25 24 23 22 21 20 19 18

◄——— GPC COUNT NO

Figure 6.8   GPC curve for polystyrene; tetrahydrofuran solvent, 22°C.

M and v must be established by calibration with monodisperse polymer standards. A calibration curve is shown in Fig. 6.9. Typically, when the data are plotted in the form $\log M$ versus v, the curve is linear over most of the range but sometimes turns up sharply at low v (high M).

Example 6.   Discuss the physical significance of an upturn in the calibration curve at low v.

Solution.   The point at which the curve begins to shoot up is set by the largest pore size in the column. All molecules that are too large to fit in those pores will pass through the column at the same rate, giving the infinite slope; that is, the technique cannot discriminate among molecules above that size. If the samples to be analyzed contain significant material above that size, it would be advisable

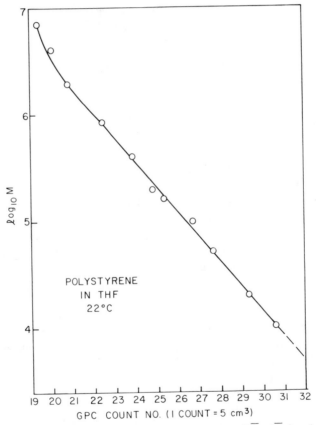

POLYSTYRENE
IN THF
22°C

GPC COUNT NO. (1 COUNT = 5 cm³)

Figure 6.9   GPC calibration; narrow-polydispersity $((\bar{M}_w/\bar{M}_n) < 1.1)$ polystyrene samples in tetrahydrofuran at 22°C.

to tack on another column (or replace the existing one) with a substrate containing larger pores to extend the calibration.

The calibration curve, strictly speaking, applies only to the particular polymer, solvent, temperature, flow rate, and column for which it was established. Change any one, and the calibration is no longer valid. Most GPC calibrations are obtained with polystyrene because it is the only polymer for which the necessary monodisperse standards are readily available. What do you do when you want to analyze another polymer? One approach was suggested by Grubisic, Rempp, and Benoit.[7] They observed that when they plotted $\log([\eta]M)$ (the product $[\eta]M$ is proportional to the hydrodynamic volume of the molecules) versus v, they obtained a single universal calibration curve for a variety of different polymers. Thus, if you measure intrinsic viscosity along with GPC elution volume for your polystyrene standards (or calculate $[\eta]$ for them with (6.16) and literature values for K and a), you can plot the universal calibration curve for your column. To back out a calibration curve for a different polymer, you would need to know K and a under the new conditions:

$$[\eta]_0 M_0 = [\eta]M \overset{6.16}{=\!=\!=} KM^{a+1} \tag{6.33}$$

$$M = \left(\frac{[\eta]_0 M_0}{K}\right)^{1/(a+1)} \tag{6.34}$$

where the subscript 0 above refers to the calibration conditions, that is, values from the universal calibration.

Balke et al.[8] have proposed a calibration method that requires only a single *polydisperse* sample of known $\bar{M}_n$ and $\bar{M}_w$. It assumes that the relation between $\log M$ and v is linear, and therefore that it can be characterized by two parameters, a slope and an intercept. Given the GPC curve and the calibration, $\bar{M}_n$ and $\bar{M}_w$ can be calculated (using techniques to be outlined below). With the GPC data on the standard, a computer program adjusts the two calibration parameters until the known $\bar{M}_n$ and $\bar{M}_w$ are obtained, in effect using the two known averages to solve for the two unknown calibration parameters.

Molecular-weight averages may be approximated directly from the GPC curve and calibration by breaking the curve into arbitrary volume increments Δv. Normally, Δv is taken as 5 cm³, the volume between "counts" on a standard machine, but smaller Δv's improve accuracy. The number of moles of polymer in a volume increment Δv is n_i:

$$n_i = \frac{c_i \Delta v}{M_i} \tag{6.35}$$

where

c_i = polymer mass concentration in the ith volume increment

M_i = molecular weight of the polymer in the ith volume increment (assumed essentially constant over small Δv)

Because $c_i = Q_i/k$ (6.32)

$$n_i = \frac{Q_i \Delta v}{k M_i}$$ (6.36)

Therefore

$$\bar{M}_n = \frac{\sum_i n_i M_i}{\sum_i n_i} = \frac{\sum_i Q_i}{\sum_i (Q_i/M_i)}$$ (6.37)

and

$$\bar{M}_w = \frac{\sum_i n_i M_i^2}{\sum_i n_i M_i} = \frac{\sum_i Q_i M_i}{\sum_i Q_i}$$ (6.38)

Note that the proportionality constant k, between concentration and GPC readout, cancels out and therefore need not be known. Q_i is normally read from the GPC curve (in any convenient units) at each count, and M_i from the calibration curve at the v corresponding to the particular count. For greater accuracy, if necessary, the integral analogs of (6.37) and (6.38) may be used, with the integrals evaluated graphically or numerically:

$$\bar{M}_n = \frac{\int_0^\infty Q \, dv}{\int_0^\infty (Q/M) \, dv}$$ (6.39)

$$\bar{M}_w = \frac{\int_0^\infty QM \, dv}{\int_0^\infty Q \, dv}$$ (6.40)

In most cases, the averages as calculated above and the qualitative information about the distribution provided by the GPC curve are all that are necessary. However, the technique *is* capable of providing the true distributions, if desired. The necessary calculations are outlined below:

$$\text{moles of polymer in volume increment } dv = \frac{c \, dv}{M} = \frac{Q}{k} \frac{dv}{M} \frac{(g/cm^3)(cm^3)}{(g/g \, mole)}$$ (6.41)

$$\frac{\text{moles of polymer}}{\text{unit of } M} = \frac{c \, dv}{M \, dM} = \frac{Q \, dv}{k M \, dM}$$ (6.42)

n = moles of polymer in a molecular-weight interval $m = \dfrac{mc\,dv}{M\,dM} = \dfrac{Qm\,dv}{kM\,dM}$

$$= \frac{mc}{2.303\,M^2}\left(\frac{dv}{d\log M}\right) = \frac{Qm}{2.303\,kM^2}\left(\frac{dv}{d\log M}\right) \tag{6.43}$$

Note:

$$\frac{dv}{d\log M} = \frac{1}{\text{slope}} \quad \text{of calibration}$$

The preceding calculations assign all the moles of polymer in a range m of the continuous GPC curve to a single spike (see Section 6.4) to calculate the correct height of the distribution.

$$N = \int_0^\infty \frac{c}{M}\,dv = \frac{1}{k}\int_0^\infty \frac{Q}{M}\,dv \tag{6.44}$$

$$\left(\frac{n}{N}\right) = \frac{(Qm/2.303\,M^2)(dv/d\log M)}{\int_0^\infty (Q/M)dv} \tag{6.45}$$

w = weight of polymer in a molecular weight interval $m = nM$

$$= \frac{mc}{2.303\cdot M}\left(\frac{dv}{d\log M}\right) = \frac{Qm}{2.303\,kM}\left(\frac{dv}{d\log M}\right) \tag{6.46}$$

W = total sample weight $= \displaystyle\int_0^\infty c\,dv = \frac{1}{k}\int_0^\infty Q\,dv$

$$= \frac{\text{area under GPC curve}}{k} \tag{6.47}$$

$$\left(\frac{w}{W}\right) = \frac{(Qm/2.303M)(dv/d\log M)}{\int_0^\infty Q\,dv} \tag{6.48}$$

Again, k cancels out. Once the distributions are known, the exact averages may be calculated by the techniques in Section 6.5. For example, insertion of (6.45) into (6.21) and (6.24) gives (6.39) and (6.40).

There are still unanswered questions about GPC, such as the effects of dispersion in the column and possible interactions of the polymer with the substrate. Some of these questions and other aspects of the technique are discussed by Abbot.[9]

REFERENCES

1. J. R. Martin, J. F. Johnson, and A. R. Cooper, *J. Macromol. Sci. – Rev. C8*, 1, 57 (1972).
2. E. A. Collins et al., *Experiments in Polymer Science*, Wiley, New York, 1973.
3. D. E. Burge, *Am. Lab.*, 9, 7, 41 (1977).
4. A. Einstein, *Ann. Phys.*, 19, 289 (1906); 34, 591 (1911).
5. F. Billmeyer, *Textbook of Polymer Science*, 2nd ed., Wiley, New York, 1971, Chapter 3.
6. M. Kurata et al., "Viscosity–Molecular Weight Relations," *Polymer Handbook*, 2nd ed., J. Brandrup and E. H. Immergut, Eds., Wiley, New York, 1975, Chapter 4-1.
7. Z. Grubisic, et al., *J. Polym. Sci.*, B-5, 753 (1967).
8. S. T. Balke et al., *Ind. Eng. Chem., Prod. Res. Dev.*, 8, 54 (1969).
9. S. D. Abbott, *Am. Lab.*, 9, 8, 41 (1977).

CHAPTER 7

Polymer Solubility and Solutions

7.1 INTRODUCTION

The thermodynamics and statistics of polymer solutions is an interesting and important branch of physical chemistry, and in itself is the subject of many good books or large sections of books. It is far beyond the scope of this chapter to attempt to cover the subject in detail. Instead, we consider topics of practical interest and try to indicate, at least qualitatively, their fundamental bases.

Three factors are of general interest:

1. What solvents will dissolve what polymers?
2. How does the polymer–solvent interaction influence the solution properties?
3. To what applications do the interesting properties of polymer solutions lead?

7.2 GENERAL RULES FOR POLYMER SOLUBILITY

Let's begin by listing some general experimental observations on the dissolution of polymers:

1. Like dissolves like; that is, polar solvents will tend to dissolve polar polymers and nonpolar solvents to dissolve nonpolar polymers. Chemical similarity of polymer and solvent is a fair indication of solubility, for example,

polyvinyl alcohol and water, $\left[\begin{array}{c} H\ H \\ -C-C- \\ H\ OH \end{array}\right]_x$ and $H{-}\overset{\displaystyle O}{\underset{\diagup\quad\diagdown}{\ }}{-}H$, or polystyrene and toluene,

$$\left[\begin{array}{c} H\ H \\ C\text{-}C \\ H\ \bigcirc \end{array}\right]_x \quad \text{and} \quad \overset{CH_3}{\bigodot}$$

2. In a given solvent at a particular temperature, the solubility of a polymer will decrease with increasing molecular weight.

3. a. Crosslinking eliminates solubility.

b. Crystallinity, in general, acts like crosslinking, but it is possible in some cases to find solvents strong enough to overcome the crystalline bonding forces and dissolve the polymer. Heating the polymer toward its crystalline melting point allows its solubility in appropriate solvents.

4. The *rate* of polymer solubility

a. increases with short branches, which loosen up the main-chain structure, allowing the solvent molecules to penetrate more easily,

b. decreases with longer branches, because the entanglement of these branches makes it harder for individual molecules to separate, and

c. decreases with increasing molecular weight.

It is important to note here that items 1, 2, and 3 are *equilibrium* phenomena and therefore describable thermodynamically (at least in principle), while 4 is a *rate* phenomenon and is governed by the rates of diffusion of polymer and solvent.

Example 1.   The polymers of ω-amino acids are termed "nylon n," where n is the number of consecutive carbon atoms in the chain. Their general formula is

$$\left[\begin{array}{c} H\ O \\ N\text{-}C\text{-}(CH_2)_{n-1} \end{array}\right]_x$$

The polymers are crystalline and do not dissolve in either water or hexane $(H_3C{-}(CH_2)_4{-}CH_3)$ at room temperature. They will, however, reach an equilibrium level of absorption when immersed in each liquid. Describe how and why water and hexane absorption varies with n.

Solution.   Water is a highly polar liquid; hexane is nonpolar. The polarity of the nylons depends on the relative proportion of polar nylon linkages $-(N{-}C)-$ with $\overset{H}{N}$ and $\overset{O}{C}$

in the chains. As n increases, the polarity of the chains decreases (they become more hydrocarbon-like), and so hexane absorption increases with n and water absorption decreases.

7.3   THE THERMODYNAMIC BASIS OF POLYMER SOLUBILITY

"To dissolve or not to dissolve. That is the question" (with apologies to W. S.). The answer is determined by the sign of the Gibbs free energy. Consider the process of mixing pure polymer and pure solvent (state 1) at constant pressure and temperature to form a solution (state 2).

$$\Delta G = \Delta H - T\Delta S \tag{7.1}$$

where ΔG = the change in Gibbs free energy in the process
ΔH = the change in enthalpy in the process
T = the absolute temperature at which the process is carried out
ΔS = the change in entropy in the process.

Only if ΔG is negative will the solution process be thermodynamically feasible. The absolute temperature must be positive, and the change in entropy for a solution process is positive, because in a solution, the molecules are in a "more random" state than in the solid. The positive product is preceded by a negative sign. Thus the third $(-T\Delta S)$ term in (7.1) favors solubility. The change in enthalpy may be either positive or negative. A positive ΔH means that the solvent and polymer "prefer their own company," that is, the pure materials are in a lower energy state, while a negative ΔH indicates that the solution is in the lower energy state. If the latter obtains, solution is assured. Negative ΔH's occur where specific interactions such as hydrogen bonds are formed between the solvent and polymer molecules. Thus if ΔH is positive, $\Delta H < T\Delta S$ if the polymer is to be soluble.

The entropy change in forming a polymer solution is quite small — orders of magnitude smaller than that which occurs when equivalent masses or volumes of two low-molecular-weight liquids are mixed. The reasons for this are illustrated qualitatively on a two-dimensional lattice model in Fig. 7.1 (such models formed the basis for early quantitative treatments of polymer solubility). In the low-molecular-weight mixture the solute molecules may be distributed randomly throughout the lattice, the only restriction being that a lattice site cannot be occupied simultaneously by two (or more) molecules. This gives rise to a large number of configurational possibilities, that is, a high entropy. In the polymer solution, however, each chain segment is confined to the lattice site adjacent to the next chain segment, greatly reducing the configurational possibilities. Note also that for a given number of chain segments (equivalent masses or volumes of polymer) the more chains they are split up into, that is, the lower the molecular

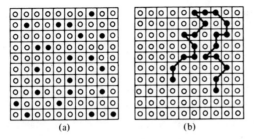

Figure 7.1   Lattice model of solubility. Solvent, open circles; solute, filled circles. (*a*) Low-molecular-weight solute; (*b*) polymeric solute.

weight, the higher will be their entropy upon solution. This explains directly the second observation, the decrease in solubility with molecular weight. But in general, for high-molecular-weight polymers, because the $T\Delta S$ term is so small, if ΔH is positive, it must be even smaller — pretty near zero, if the polymer is to be soluble.

7.4   THE SOLUBILITY PARAMETER

How can ΔH be estimated? Well, for *regular* solutions (those in which solute and solvent *do not* form specific interactions) the change in *internal energy* upon solution is given by

$$\Delta H \cong \Delta E = \phi_1\phi_2(\delta_1 - \delta_2)^2 \left(\frac{cal}{cm^3 \, solution}\right) \qquad (7.2)$$

where ΔE = the change in internal energy in the solution process
ϕ = volume fractions
δ = *solubility parameters*

The subscripts 1 and 2 usually refer to solute (polymer) and solvent, respectively. The *solubility parameter* is defined as follows:

$$\delta = (CED)^{1/2} = \left(\frac{\Delta E_v}{v}\right)^{1/2} \; [=] \; \left(\frac{cal}{cm^3}\right)^{1/2} = hildebrand \qquad (7.3)$$

where (CED) = *cohesive energy density*, a measure of the strength of the inter-molecular forces holding the molecules together in the liquid state
ΔE_v = molar change in internal energy on vaporization, cal/g mole
v = molar volume of liquid, cm³/g mole

Now, for a process that occurs at constant volume and constant pressure, the changes in internal energy and enthalpy are equal. Since the change in volume on solution is usually quite small, this is a good approximation for the dissolution of polymers under most conditions; so (7.2) provides a means of estimating enthalpies of solution if the solubility parameters of the polymer and solvent are known.

Note that regardless of the magnitudes of δ_1 and δ_2 (they are always positive), the predicted ΔH is always positive, so, (7.2) applies only in the absence of specific interactions that lead to negative ΔH's. Inspection of (7.2) also reveals that ΔH is minimized and the tendency toward solubility maximized by matching the solubility parameters as closely as possible. As a rough rule of thumb

$$|\delta_1 - \delta_2| < 0.5 \qquad \text{for solubility} \qquad (7.4)$$

Measuring the solubility parameter of a low-molecular-weight solvent is no problem. Polymers, on the other hand, degrade long before reaching their vaporization temperatures, making it impossible to evaluate ΔE_v directly. Fortunately, there is a way around this impasse. The greatest tendency of a polymer to dissolve occurs when its solubility parameter matches that of the solvent. If the polymer is crosslinked lightly, it cannot dissolve but only swell. The maximum swelling will be observed when the polymer and solvent solubility parameters are matched. So polymer solubility parameters are determined by soaking lightly crosslinked samples in a series of solvents of known solubility parameters. The value of the solvent at which maximum swelling is observed is taken as the solubility parameter of the polymer (Fig. 7.2). Solubility parameters of solvent mixtures can be readily calculated:

$$\delta_{\text{mixture}} = \frac{x_1 v_1 \delta_1 + x_2 v_2 \delta_2}{x_1 v_1 + x_2 v_2} \qquad (7.5)$$

where x = mole fraction

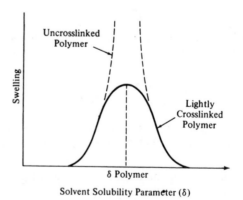

Figure 7.2   Determination of polymer solubility parameter by swelling lightly crosslinked samples in a series of solvents.

In the absence of specific data on solvents, a method has been published for estimating both the solubility parameters and molar volumes of liquids.[1]

While the solubility parameter concept has proved useful, there are unfortunately many exceptions to (7.4). Polymer solubility is too complex a phenomenon to be described quantitatively with a single parameter. Several techniques that supplement solubility parameters with quantitative information on hydrogen bonding and dipole moments[2,3] have been proposed. One of the more useful of these is discussed in the next section.

7.5   HANSEN'S THREE-DIMENSIONAL SOLUBILITY PARAMETER

According to Hansen, the total change in internal energy on vaporization ΔE_v may be considered the sum of three individual contributions, one due to hydrogen bonds ΔE_h, another due to permanent dipole interactions ΔE_p, and a third from dispersion (van der Waals or London) forces ΔE_d:

$$\Delta E_v = \Delta E_d + \Delta E_p + \Delta E_h \tag{7.6}$$

Dividing by the molar volume v gives

$$\frac{\Delta E_v}{v} = \frac{\Delta E_d}{v} + \frac{\Delta E_p}{v} + \frac{\Delta E_h}{v} \tag{7.7}$$

or

$$\delta^2 = \delta_d^2 + \delta_p^2 + \delta_h^2 \tag{7.8}$$

where

$$\delta_i = \left(\frac{\Delta E_i}{v}\right)^{1/2} \tag{7.9}$$

The solubility parameter δ may be thought of as a vector in a three-dimensional $\delta_p, \delta_d, \delta_h$ space. Equation (7.8) gives the magnitude of the vector in terms of its components. A solvent, therefore, with given values of δ_p, δ_d, and δ_h is represented as a point in space, with δ being the vector from the origin to this point.

A polymer is also characterized by values of δ_p, δ_d, and δ_h. Furthermore, it has been found on a purely empirical basis that if δ_d is plotted on a scale twice the size as that used for δ_p and δ_h, all solvents that will disolve that polymer fall within a sphere of radius R surrounding the point $(\delta_p, \delta_d, \delta_h)$ for the polymer. Figure 7.3 shows the solubility sphere for polystyrene ($\delta_d = 8.6$, $\delta_p = 3.0$, $\delta_h = 2.0$, $R = 3.5$).[4] It will be noted that parts of the polystyrene sphere lie outside the first octant. The physical significance of these areas is questionable, at best.

Values of the individual components δ_p, δ_d, and δ_h have been developed from measured δ values, theoretical calculations, studies on model compounds, and plenty of computer fitting. They are extensively tabulated for solvents.[4-8]

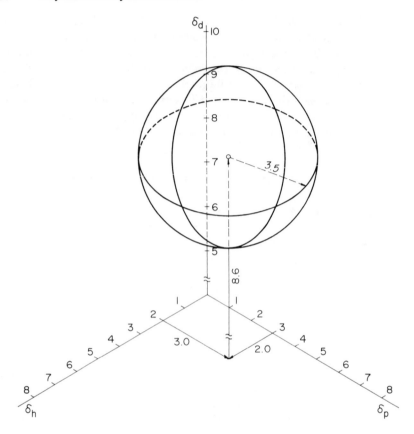

Figure 7.3   The Hansen solubility sphere for polystyrene ($\delta_d = 8.6$, $\delta_p = 3.0$, $\delta_h = 2.0$, $R = 3.5$).[4]

They, along with R, are less readily available for polymers but have been published.[4,5]

Despite its empirical nature, the three-dimensional solubility parameter has proved of great practical utility, particularly in the paint industry, where the economic choice of solvents is of critical importance. It is capable of explaining those cases in which solvent and polymer δ's are almost perfectly matched, and yet the polymer will not dissolve (the δ vectors have the same magnitudes, but different directions), or where two nonsolvents can be mixed to form a solvent (the solvent components lie on opposite sides outside the sphere, the mixture within). Inorganic pigments may also be characterized by δ vectors. Pigments whose δ vectors closely match those of a solvent tend to form stable suspensions in that solvent.

Example 2. A polymer has a solubility parameter $\delta = 9.95$ ($\delta_p = 7.0$, $\delta_d = 5.0$, $\delta_h = 5.0$) and a solubility sphere of radius $R = 3.0$ (all numbers in hildebrands). Will a solvent with $\delta = 10$ ($\delta_p = 8$, $\delta_d = 6$, $\delta_h = 0$) dissolve it?

Solution. No. The solvent point lies in the $\delta_p - \delta_d$ plane (i.e., $\delta_h = 0$). The closest approach the polymer solubility sphere makes to this plane is $5.0 - 3.0 = 2$. Thus, despite nearly identical δ's, the solvent will not dissolve the polymer.

7.6 PROPERTIES OF DILUTE SOLUTIONS

Well, now the polymer is in solution. Let's assume it's a fairly dilute solution so that we don't have to worry about too many entanglements between the molecules. In a "good" solvent (one whose solubility parameter closely matches that of the polymer) the secondary forces between polymer segments and solvent molecules are strong, and the polymer molecules will assume a spread-out conformation in solution. In a "poor" solvent the attractive forces between the segments of the polymer chain are greater than those between the chain segments and the solvent; that is, the chain segments "prefer their own company," and the chain will ball up tightly (Fig. 7.4).

Imagine a polymer in a "good" solvent, for example, polystyrene ($\delta = 9.3$) in chloroform ($\delta = 9.2$). A nonsolvent is now added, say, methanol ($\delta = 14.5$). (Despite the large difference in solubility parameters, chloroform and methanol are mutually soluble in all proportions because of the large ΔS of solution for low-molecular-weight compounds.) Ultimately, a point is reached where the mixed solvent becomes too "poor" to sustain solution, and the polymer

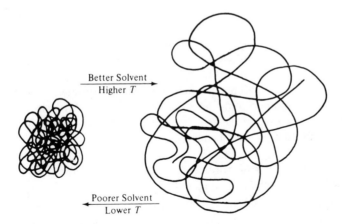

Figure 7.4 The effects of solvent power and temperature on a polymer molecule in solution.

precipitates out as the attractive forces between polymer segments becomes much greater than those between polymer and solvent. At some point the polymer teeters on the brink of solubility when $\Delta G = 0$ and $\Delta H = T\Delta S$. This point obviously depends on the temperature, polymer molecular weight (mainly through its influence on ΔS), and the polymer–solvent system (mainly through its influence on ΔH). Adjusting either the temperature or the polymer–solvent system allows fractionation of the polymer according to molecular weight, as successively smaller molecules precipitate upon lowering the temperature or going to poorer solvent. In the limit of infinite molecular weight (the minimum possible ΔS), the situation in which $\Delta H = T\Delta S$ is known as the "θ condition." Under these conditions the polymer–solvent and polymer–polymer inter-actions are equal, and the solution behaves in a so-called ideal fashion, with the second virial coefficient equal to 0, and so on. For a given polymer, the θ condition can be reached at a fixed temperature by adjusting the solvent to give a "θ solvent," or with a particular solvent by adjusting the temperature to reach the θ or "Flory" temperature. Any actual polymer will still be soluble under θ conditions, of course, because of its lower-than-infinite molecular weight and consequently larger ΔS.

Getting back to our example, we might ask what happens to the viscosity of the solution upon going from a good solvent to a poor solvent (to make a fair comparison, imagine that as nonsolvent is added, an equivalent amount of good solvent is removed, maintaining a constant polymer concentration). This question can be answered qualitatively by imagining the polymer molecules in solution to be rigid spheres (they aren't) and applying the Einstein relation equation (6.10) (further assuming that the viscosities of the low-molecular-weight liquids are comparable, i.e., that η_s doesn't change). Going from good to poor solvent, the polymer molecules "ball up," giving a smaller effective ϕ, *lowering* the solution viscosity. Thus solution viscosity can be controlled by adjusting solvent power. This fact is of great importance to the surface-coatings industry. For example, in formulating a lacquer (a solution of polymer in solvent plus some pigment, which dries only by solvent evaporation), the viscosity for optimum spraying or brushing characteristics might be obtained with a mixed solvent, with the poorer component the more volatile. After application, the poorer component evaporates first, leaving a high viscosity and, therefore, sag- and run-resistant film on the substrate.

Example 3.   Indicate how solvent "power" ("good" solvent versus "poor" solvent) will influence the following:

a.   The design of a stirred solution polymerization reactor.

b.   The intrinsic viscosity of a polymer sample at a particular T.

c.   The molecular weight of a polymer sample as determined by membrane osmometry.

Solution

a. As long as the polymer remains in solution, a "poor" solvent, with its lower viscosity, will permit better agitation or require a less powerful stirrer motor.

b. The measured viscosities of solutions will be less in a "poor" solvent than in a "good" one, leading to lower intrinsic viscosities.

c. If the membrane were ideal, solvent power would make no difference, as osmometry measures only the number (moles) of solute per unit volume, regardless of geometry. Some of the smallest particles can "sneak" through any real membrane, however. The "poor" solvent, causing the molecules to "ball up" tightly, will allow passage of more molecules through the membrane. This will lower the observed osmotic pressure, causing the calculated $\bar{M}_n$ to err on the high side.

Consider now the effects of temperature on the viscosity of a solution of polymer in a relatively poor solvent. As is the case with all simple liquids, the solvent viscosity η_s decreases with increasing temperature. Increasing temperature, however, imparts more thermal energy to the segments of the polymer molecules, causing the molecules to spread out and assume a larger effective ϕ in solution. Thus the effects of temperature on η_s and ϕ tend to compensate, giving a solution that has a much smaller change in viscosity with temperature than that of the solvent alone. In fact, the additives that are used to produce the so-called multiviscosity (10W-40) motor oils are nothing more than polymers, with compositions adjusted so that the base oil is a relatively poor solvent at the lowest operating temperatures. As the engine heats up, the polymer molecules uncoil, providing a much greater resistance to "thinning" than is possible from oil alone.

Viscosities (at low shear rates) of dilute solutions of polymers with known molecular weights may be calculated using (6.15) and (6.16), reversing the procedure for obtaining molecular weights from viscosity measurements.

7.7 POLYMER–POLYMER COMMON SOLVENT SYSTEMS

We have discussed the small increase in entropy that arises when a high-molecular-weight polymer is dissolved in a solvent. If we apply the same reasoning to the dissolution of one high-molecular-weight polymer in another, the ΔS of solution will be even smaller. For this reason, the true solubility of one polymer in another is relatively rare. Polymer–polymer miscibility has been extensively reviewed.[9] Even when a common solvent is added (one that is infinitely soluble with each polymer alone), the two polymers usually cannot coexist in a homogenous phase beyond a few percent concentration. A schematic phase diagram for such a system is shown in Fig. 7.5. Beyond a few percent of polymer (the

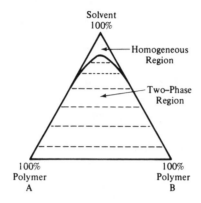

Figure 7.5 Typical ternary phase diagram for a polymer–polymer common solvent system. The dotted tie lines connect compositions of phases in equilibrium.

exact value depends on the chemical nature of the polymers and solvent and the molecular weights of the polymers), two phases in equilibrium are formed, each phase containing a nearly pure polymer. These results extend to more than two polymers. In general, each polymer will coexist in a separate phase.

7.8   CONCENTRATED SOLUTIONS–PLASTICIZERS

Up to this point we have considered relatively dilute polymer solutions. Now let's look at the other end of the spectrum, where the polymer is the major constituent of the solution. A pure amorphous polymer consists of a tangled mass of polymer chains. The ease with which this mass can deform depends on the ability of the polymer chains to untangle and slip past one another. One way of increasing this ability is to raise the temperature, as we have seen. Another is to add a low-molecular-weight liquid, an *external plasticizer,* to the polymer. By forming secondary bonds to the polymer molecules and spreading them apart, the plasticizer reduces the polymer–polymer secondary bonding and provides more room for the polymer molecules to move around, giving a softer, more easily deformable mass.

Example 4. If you have ever heard baseball announcers comment that "On these humid nights, the ball just isn't hit as hard" or "A pitcher throws a heavier ball on these humid nights," you may have wondered if there was a scientific basis for these statements. They cannot be justified in terms of the properties of the humid air (fluid dynamics). Can you justify them in terms of the baseball properties? A baseball is made largely of tightly wound wool yarn. Wool, a natural polymer, is similar, physically and chemically, to nylons (see Example 1).

Solution.   The polar nylon linkages hydrogen bond to water. The higher the humidity, the greater the equilibrium moisture content of the wool, accounting for the heavier ball. The absorbed water also acts as an external plasticizer for the wool, softening it and decreasing its resiliency.

Quantitative confirmation of Example 4 was provided by the noted technical journal, *Sports Illustrated*, which reported on the effects of moisture on the mass and resilience of baseballs:[10]

> . . . balls baked in a 212°F oven for 24 hours lost 12 grams and gained 5.8% in vigor . . . .

> . . . 24 hours in a steam-filled moist room caused them to pick up 11 grams and lose 17.4% of their bounce . . . .

A similar reversible plasticization of fibers by heat and moisture is applied in the process of steam ironing fabrics.

A "poor" (from the thermodynamic standpoint) plasticizer should be more effective than a "good" one, giving a lower viscosity at a given level (smaller ϕ, fewer entanglements), but since it is less strongly bound to the polymer, it will have a greater tendency to exude out over a period of time, leaving the polymer mass stiffer. This was a familiar problem with early shower curtains, for example. Thus a balance must be struck between plasticizer efficiency and permanence. Also, from the standpoint of permanence, it is necessary that the plasticizer have low volatility. Plasticizers, therefore, generally have higher molecular weights than solvents but are still well below the high-polymer range in this respect.

A polymer may be *internally plasticized* by random copolymerization with a monomer whose homopolymer is very soft (has a low T_g – see Chapter 8). There are obviously never any permanence problems when this is done. The composition of the copolymer can be adjusted to give the desired properties at a particular temperature. Above this temperature the copolymer will be softer than intended and below, harder. With external plasticizers, the same sort of temperature compensation as in the multiviscosity motor oils is obtained, giving materials that maintain the desired flexibility over a broader temperature range.

The most common externally plasticized polymer is polyvinyl chloride (PVC)

$$\left[\begin{array}{c} \text{H} \ \ \text{H} \\ | \ \ \ | \\ \text{C--C} \\ | \ \ \ | \\ \text{H} \ \ \text{Cl} \end{array}\right]_x$$. A typical plasticizer for PVC is dioctyl phthalate (DOP) (actually

di-2 ethylhexyl phthalate), the di-ester of phthalic acid and iso-octyl alcohol

$$
\begin{array}{c}
\text{CH}_3 \\
|\\
\text{HCH} \\
\overset{O}{\overset{\|}{C}}\text{-O-}\overset{H}{\underset{H}{C}}\text{-}\overset{H}{\underset{H}{C}}\text{-(CH}_2)_3\text{CH}_3 \\
\overset{}{C}\text{-O-}\overset{H}{\underset{H}{C}}\text{-}\overset{H}{\underset{H}{C}}\text{-(CH}_2)_3\text{CH}_3 \\
\overset{\|}{O}\quad \text{HCH} \\
|\\
\text{CH}_3
\end{array}
$$

Phosphates, tricresyl phosphate (TCP), for example, are used (as are chlorinated waxes) as plasticizers that impart fire resistance. So-called polymeric plasticizers, polymers with molecular weights on the order of 1000, provide low volatility and good permanence. It might be noted that a low-molecular-weight fraction in a pure polymer behaves as a plasticizer, often in an undesirable fashion.

Epoxidized plasticizers are becoming increasingly important. In addition to plasticizing PVC, these compounds also perform the important function of stabilizing the resin. When PVC degrades, HCl is given off, and the HCl catalyzes further degradation. Not only that, it attacks metallic molding machines, molds, and extruders. The epoxy plasticizers are made from unsaturated oils:

$$
\sim\!\!\sim\!\!C\overset{H}{=}\!C\overset{H}{\sim}\!\!\sim + \ H_2O_2 \ \xrightarrow{\text{catalyst}} \ \sim\!\!\sim\!\!\underset{\underset{O}{\diagdown\diagup}}{C}\overset{H}{-}\!\underset{}{C}\overset{H}{\sim}\!\!\sim \ + \ H_2O
$$

Oxirane ring

The *oxirane* rings soak up HCl and minimize further degradation (this is an admittedly oversimplified view of a very complex phenomenon):[11]

$$
\sim\!\!\sim\!\!\underset{\underset{O}{\diagdown\diagup}}{C}\overset{H}{-}\!\underset{}{C}\overset{H}{\sim}\!\!\sim \ + \ HCl \ \rightarrow \ \sim\!\!\sim\!\!\underset{OH}{\overset{H}{C}}\!-\!\underset{Cl}{\overset{H}{C}}\!\!\sim\!\!\sim
$$

Unplasticized (or nearly so) PVC is a rigid material used for pipe and fittings, among other things. The properties of plasticized PVC vary considerably depending on the plasticizer level. It is familiar as a gasketing material, leather-like upholstery material, wire and cable covering, shower curtains, and so forth.

Plastisols are an interesting and useful technological application of plasticized PVC. A typical formulation might consist of 100 parts DOP phr (per hundred parts resin) plus some stabilizers, pigments etc. Although the PVC is thermo-dynamically soluble in the DOP at room temperature, the rate of dissolution is extremely slow. So initially the plastisol is a milky suspension of finely divided resin particles in the plasticizer. As a suspension rather than a solution, the viscosity is of the order of magnitude of that of the plasticizer itself, and it can be applied to substrates or molds by brushing, dipping, rolling, etc. When heated to about 350°F, the increased thermal agitation of the polymer molecules speeds up the solution process greatly. If there is no filler or pigment present, the solution process can be observed as the plastisol becomes transparent. When the plastisol is cooled back to room temperature, the viscosity of the solution is so high that for all practical purposes, it may be considered a flexible solid. Examples of plastisols include doll "skin" and the covering of wire dish racks. The viscosity of the initial suspension may be lowered by incorporating a volatile organic diluent for the plasticizer, giving an *organsol*, or by whipping in water to form a *hydrosol*. In both cases the diluent vaporizes upon heating.

REFERENCES

1. R. F. Fedors, *Polym. Eng. Sci.,* **14**, 147 (1974); **14**, 472 (1974).
2. A. Beerbower, L. A. Kaye, and D. A. Pattison, *Chem. Eng.,* **74**, 26, 118 (1967).
3. R. B. Seymour, *Mod. Plast.,* **48**, 10, 150 (1971).
4. C. M. Hansen, *The Three Dimensional Solubility Parameter and Solvent Diffusion Coefficient,* Danish Technical Press, Copenhagen, 1967.
5. C. M. Hansen, *J. Paint Technol.,* **39**, 505, 104 (1967).
6. Hansen, C. M. and A. Beerbower, "Solubility Parameters," *Encyclopedia of Chemical Technology,* Suppl. Vol., 2nd ed., Wiley, New York, 1971.
7. C. M. Hansen, *Ind. Eng. Chem., Prod. Res. Dev.,* **8**, 2 (1969).
8. J. Brandrup and E. H. Immergut, Eds., *Polymer Handbook,* 2nd ed., Wiley-Interscience, New York, 1975.
9. O. Olabisi, L. M. Robeson, and M. T. Shaw, *Polymer-Polymer Miscibility,* Academic Press, New York, 1979.
10. *Sports Illus.,* Time, July 20, 1970, p. 22.
11. D. F. Anderson and D. A. McKenzie, *J. Polym. Sci.,* **A-1**, 8, 2905 (1970).

CHAPTER 8

Transitions in
Polymers

8.1 THE GLASS TRANSITION

It has long been known that amorphous polymers can exhibit two distinctly different types of mechanical behavior. Some, like polymethyl methacrylate (Lucite, Plexiglas) and polystyrene, are hard, rigid, *glassy* plastics at room temperature, while others, for example, polybutadiene, polyethyl acrylate and polyisoprene, are soft, flexible, rubbery materials. If, however, polymethyl methacrylate and polystyrene are heated to about 125°C, they exhibit typical rubbery properties, and when a rubber ball is cooled in liquid nitrogen, it becomes rigid and glassy, and shatters when an attempt is made to bounce it. So there is some temperature, or narrow range of temperatures, below which an amorphous polymer is in a glassy state, and above which it is rubbery. This temperature is known as the *glass-transition temperature T_g*. The glass-transition temperature is a property of the polymer, and whether the polymer has glassy or rubbery properties depends on whether its application temperature is above or below its glass transition temperature.

8.2 MOLECULAR MOTIONS IN AN AMORPHOUS POLYMER

In order to understand the molecular basis for the glass transition, the various molecular motions occurring in an amorphous polymer mass may be broken into four categories:

1. translational motion of entire molecules, which permits flow;
2. cooperative wriggling and jumping of segments of molecules approximately 40 to 50 carbon atoms in length, permitting flexing and uncoiling;

88

3.   motions of a few atoms along the main chain (five or six, or so) or of side groups on the main chains;

4.   vibrations of atoms about equilibrium positions, as occur in crystal lattices, except that the atomic centers are not in a regular arrangement in an amorphous polymer.

Motions 1–4 above are arranged in order of decreasing activation energy; that is, smaller amounts of thermal energy (kT) are required to produce them. The glass-transition temperature is thought to be that temperature at which motions 1 and 2 are pretty much "frozen out" and there is only sufficient energy available for motions of types 3 and 4. Of course, not all molecules possess the same energies at a given temperature. The molecular energies follow a Boltzmann distribution, and even below T_g there will be occasional type 2 and even type 1 motions, which can manifest themselves over extremely long periods of time.

8.3   DETERMINATION OF T_g

How is the glass transition studied? A common method is to observe the variation of some thermodynamic property with T, for example, the specific volume, as shown in Fig. 8.1. Note that the slope of the v versus T plot increases above the glass-transition temperature.

The value of T_g determined in this fashion will vary slightly with the rate of cooling or heating. This reflects the fact that long, entangled polymer chains

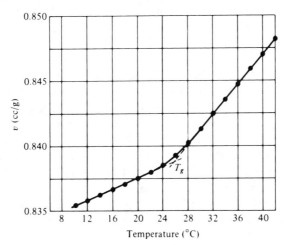

Figure 8.1   Specific volume versus temperature for polyvinyl acetate.[1]

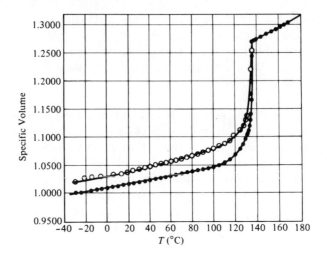

Figure 8.2   Specific volume–temperature relations for linear polyethylene (Marlex-50). Specimen slowly cooled from melt to room temperature prior to fusion, open circles; specimen crystallized at 130°C for 40 days and then cooled to room temperature prior to fusion, closed circles.[2]

cannot respond instantaneously to changes in temperature and illustrates the difficulty in making thermodynamic measurements on polymers. It often takes extremely long to reach equilibrium, if indeed it is ever reached, and it is difficult to be sure if and when it is reached. Strictly speaking, the glass-transition temperature should be defined in terms of equilibrium properties – or at least those measured with very low rates of temperature change. Also, there is never observed a sharp "break" in the property, but this is of little practical importance because T_g can always be established within a couple of degrees by extrapolation of the linear regions. Other properties, such as refractive index, may also be used to establish T_g.

In contrast to a change in *slope* at the glass transition, a thermodynamic property such as specific volume exhibits a *discontinuity* with temperature at the crystalline melting point in polymers as in other materials (Fig. 8.2). The glass transition is therefore known as a *second-order* thermodynamic transition (v versus T continuous, dv/dT versus T discontinuous) in contrast to a first-order transition such as the melting point (v versus T discontinuous). There is still considerable argument as to whether T_g is a true second-order transition, but this is of little practical consequence. Since all polymers have at least some amorphous material (they can't be 100% crystalline), they all have a T_g; but not all polymers have a crystalline melting point – they can't if they don't have any crystallinity.

Transitions in polymers are rapidly and conveniently studied using DTA

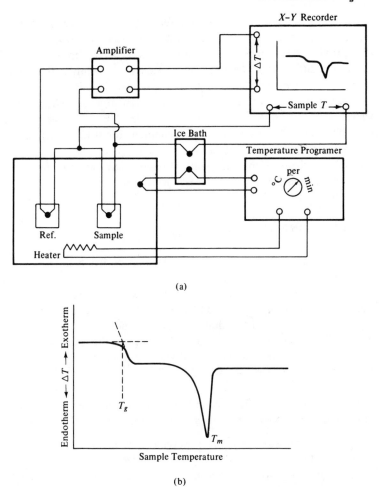

Figure 8.3 Differential thermal analysis. (*a*) Schematic of apparatus; (*b*) DTA curve.

(differential thermal analysis) and DSC (differential scanning calorimetry). In DTA small samples of the polymer and an inert reference substance (one that undergoes no transition in the temperature range of interest) are mounted in a block with thermocouples to monitor temperatures (Fig. 8.3). The thermodynamic property monitored here is the enthalpy. The block is heated at a constant rate, and the *difference* between the sample and reference temperatures is plotted versus the sample temperature. At T_g, $C_p = (\partial H/\partial T)_p$ increases (the slope of an H versus T plot increases) dropping the sample temperature below its previous level relative to the reference. At T_m large amounts of heat are required to melt the crystals at constant T, and the ΔT shows a sharp dip.

DSC provides the same sort of information. Here, however, the device measures the differential power input necessary to maintain the sample and reference at the *same* temperature as they are heated. This technique is somewhat more amenable to quantitative measurements of ΔH_m, and other properties.

Both DTA and DSC are also useful techniques for studying oxidation (exothermic) and degradation (endothermic) reactions.

Dynamic mechanical measurements (Chapter 18) can also provide useful information on transitions.

8.4   FACTORS INFLUENCING T_g[3]

In general, the glass-transition temperature depends on five factors:

1.   The *free volume* of the polymer v_f. Free volume is the volume of the polymer mass not actually occupied by the molecules themselves; that is, $v_f = v - v_s$, where v is the specific volume of the polymer mass and v_s is the volume of the solidly packed molecules. The higher the v_f, the more room the molecules will have in which to move around and the lower the T_g. It has been estimated that for all polymers $v_f/v = 0.025$ at T_g.

Example 1.   Glass-transition temperatures have been observed to increase at pressures of several thousand pounds per square inch. Why?

Solution.   High pressures compress polymers, reducing v. Since v_s doesn't change appreciably, v_f is reduced.

2.   The attractive forces between the molecules. The more strongly they are bound together, the greater the thermal energy required to produce motion. Since the solubility parameter δ is a measure of intermolecular forces, T_g increases with δ. Polyacrylonitrile $\left(-\left[\begin{array}{c} \text{H} \;\; \text{H} \\ \text{C--C} \\ \text{H} \;\; \text{C} \end{array}\right]_x\!-\right)$, because of extremely strong

secondary bonding between chains, has a T_g higher than its degradation temperature.

3.   The internal mobility of the chains, that is, their freedom to rotate about bonds. Figure 8.4 shows potential energy as a function of rotation angle about a bond in a polymer chain. The minimum energy configuration, arbitrarily chosen as $\theta = 0$, is the position where the largest substituents, the rest of the

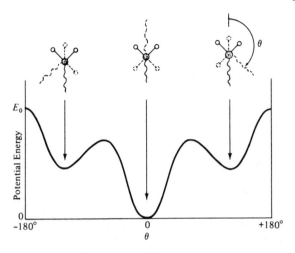

Figure 8.4 Rotation about a bond in a polymer chain backbone, viewed along the bond. The dotted substituents are on the rear carbon atom.

chain, are as far away from each other as possible. As the bond is rotated, the substituent groups are brought into juxtaposition, and energy is required to "push them over the hump." The maximum energy is needed to get the two chain substituents past one another, and this energy must be available if complete rotation is to be obtained. Table 8.1 shows how T_g increases with E_0 for a series of polymers with approximately the same δ. Note how the ether oxygen "swivel" in the silicone chain permits very free rotation.

Table 8.1[3]

Polymer		δ	$T_g, °C$	E_0, kcal/mole
Silicone rubber	$\left[\begin{array}{c}CH_3\\ \mid\\ Si-O\\ \mid\\ CH_3\end{array}\right]_x$	7.3	-120	~ 0
Polyethylene	$\left[\begin{array}{c}H\ H\\ \mid\ \mid\\ C-C\\ \mid\ \mid\\ H\ H\end{array}\right]_x$	7.9	-85	3.3
Polytetrafluoroethylene	$\left[\begin{array}{c}F\ F\\ \mid\ \mid\\ C-C\\ \mid\ \mid\\ F\ F\end{array}\right]_x$	6.2	> 20	4.7

Example 2.   Poly α-methylstyrene has a higher T_g than polystyrene. Why?

$$\begin{array}{cc}
\text{Polystyrene} & \text{Poly } \alpha\text{–methylstyrene}
\end{array}$$

Solution.   The methyl group introduces extra steric hindrance to rotation, giving a higher E_0.

4.   The "stiffness" of the chains. Chains that have difficulty coiling and folding will have higher T_g's. This stiffness usually goes hand in hand with high E_0, so it is difficult to separate the effects of 3 and 4.

Chains with parallel bonds in the backbone (ladder-type polymers), for example, polyimides (Chapter 2, Example 4R), and those with highly aromatic backbones have extremely stiff chains and, therefore, high T_g's. This makes them mechanically useful at elevated temperatures, but also very difficult to process.

5.   The chain length. As do many mechanical properties of polymers, the glass transition temperature varies according to the empirical relation

$$T_g = T_g^\infty - \frac{C}{x} \tag{8.1}$$

where C is a constant for the particular polymer and T_g^∞ is the asymptotic value of the glass-transition temperature at infinite chain length. In most cases, for $x > 500$ (the range of commercial interest for most polymers), $T_g \cong T_g^\infty$.

Example 3.   Measurements such as those described above show that an external plasticizer softens a polymer by reducing its glass-transition temperature. Explain.

Solution.   The plasticizer molecules "pry apart" the polymer chains, in essence increasing the free volume available to the chains (although not truly "free," the small plasticizer molecules interfere with chain motions much less than would other chains). Also, by forming secondary bonds with the polymer chains, the plasticizer molecules reduce the bonding forces between the chains themselves.

8.5 T_g'S OF COPOLYMERS

The glass-transition temperatures for random copolymers vary monotonically with composition between those of the homopolymers. They can be approximated fairly well from a knowledge of the T_g's of the homopolymers with the empirical relation

$$\frac{1}{T_g} = \frac{\omega_1}{T_{g1}} + \frac{\omega_2}{T_{g2}} \tag{8.2}$$

where the ω's are weight fractions of the monomers in the copolymer. This relation forms the basis for a method of estimating the T_g's of highly crystalline polymers, in which the properties of the small amount of amorphous material are masked by the majority of crystalline material present. If it is possible to produce a series of random copolymers in which the randomness prevents crystallization over a certain composition range, (8.2) can be used to extrapolate to $\omega_1 = 0$ and $\omega_1 = 1$, giving the T_g's of the homopolymers. This method is open to question because it assumes that the presence of major amounts of crystallinity does not restrict the molecular response in the amorphous regions. In fact, the T_g's of highly crystalline polymers (polyethylene, in particular) are still open to debate.

8.6 THE THERMODYNAMICS OF MELTING

In polymers the crystalline melting point T_m is a phase change similar to that observed in low-molecular-weight organic compounds, metals, and ceramics.

The (Gibbs) free energy of melting is given by

$$\Delta G_m = \Delta H_m - T\Delta S_m \tag{8.3}$$

At the crystalline melting point T_m, $\Delta G_m = 0$; so

$$T_m = \frac{\Delta H_m}{\Delta S_m} \tag{8.4}$$

Now ΔH_m is the energy needed to overcome the crystalline bonding forces at constant T and P, and is essentially independent of chain length for high polymers. For a given mass or volume of polymer, however, the shorter the chains, the more "randomized" they become upon melting, giving a higher ΔS_m. (For a more detailed description of this in connection with solutions, see Chapter 7.) Thus, the crystalline melting point decreases with decreasing chain length, and in a polydisperse polymer the distribution of chain lengths gives a distribution of melting points or "rounding off" noted in Fig. 8.2.

Equation (8.4) also indicates that chains that are strongly bound in the crystal lattice, that is, have a high ΔH_m, will have a high T_m, as expected. Also, the stiffer and less mobile chains, those that can "randomize" less upon melting and therefore have a low ΔS_m, will tend to have higher T_m's.

Example 4. Discuss how the crystalline melting point varies with n in the "nylon n" series (Chapter 7, Example 1).

Solution.   Increasing values of n dilute the nylon linkages that are responsible for interchain hydrogen bonding and thus lower the crystalline melting point. As n goes to infinity, the structure approaches that of linear polyethylene. This represents the asymptotic minimum T_m, with the chains held together only by van der Waal's forces.

Table 8.2 illustrates the variation in T_m and some other properties with n for some commerical members of the series.

Table 8.2   Variation of properties with n for nylon n's

n	$T_m, °C$	ρ, g/cm^3	Tensile strength, psi	Water absorption, % in 24 hr
6	216	1.14	12,000	1.7
11	185	1.04	8,000	0.3
12	177	1.02	7,500	0.25
∞ (PE)	135	0.97	5.500	nil

Example 5.   Consider the following classes of linear, aliphatic polymers:

Polyurethanes		Polyamides		Polyureas
T_m	<	T_m	<	T_m

For given values of n and x the crystalline melting points increase from left to right, as indicated. Explain.

Solution.   The polyurethane chains contain the —O— "swivel"; thus they are the most flexible and have the largest ΔS_m and lowest T_m. Hydrogen bonding, and thus ΔH_m, is roughly comparable in the polyurethanes and polyamides. The polyureas and polyamides should have chains of comparable flexibility (no swivel), but with the extra —N—, the polyureas form stronger or more extensive hydrogen bonds and therefore have a higher ΔH_m than the polyamides.

The crystalline melting point also increases a bit with the degree of crystallinity of a polymer. For example, low-density polyethylene (about 65% crystalline) has a T_m of about 115°C, whereas high-density polyethylene (about 95% crystalline) melts at about 135°C. This can be explained by treating the amorphous material as an impurity. It is well known that the introduction of an impurity lowers the melting point of common materials. In a similar fashion, greater amounts of noncrystalline "impurities" lower the crystalline melting point of a polymer.

Since polymer chains are largely immobilized below T_g, if they are cooled rapidly through T_m to below T_g, it is sometimes possible to obtain normally crystalline polymers in a metastable amorphous state which will persist indefinitely. When "annealed" above T_g (and below T_m), they will crystallize, as the chains gain the mobility necessary to pack into a lattice.

Example 6. Polyethylene terephthalate (Mylar®, Dacron®) is cooled rapidly from 300°C (state 1) to room temperature. The resulting material is rigid and perfectly transparent (state 2). The sample is then heated to 100°C (state 3) and maintained at that temperature, during which time it gradually becomes translucent (state 4). It is then cooled down to room temperature and is again found to be rigid, but is now translucent rather than transparent (state 5). For this polymer, $T_m = 267°C$ and $T_g = 69°C$. Sketch a general specific volume versus temperature curve for a crystallizable polymer illustrating T_g and T_m, and show the locations of states 1–5 for the sample above.

Solution. Figure 8.5 illustrates the general v versus T curve for a crystallizable polymer. The dotted upper portion represents the metastable amorphous material obtainable by rapid cooling. The history above is shown on the diagram. The metastable amorphous material (transparent, state 2) obtained by rapid cooling to below T_g crystallizes upon annealing between T_g and T_m (state 3 to state 4).

The greater mobility of chains crystallized just below T_m accounts for their higher degree of crystallinity observed in Fig. 8.2.

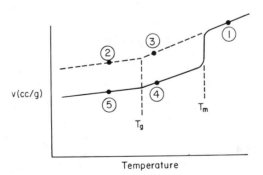

Figure 8.5   Specific volume–temperature relation for crystallizable polymer. Numbers apply to Example 6.

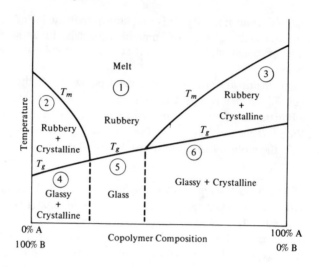

Figure 8.6   Phase diagram for a random copolymer system.

8.7   THE INFLUENCE OF COPOLYMERIZATION ON PROPERTIES

The influence of random copolymerization on T_m and T_g is interesting and technologically important. Occasionally, if two repeating units are similar enough sterically to fit into the same crystal lattice, random copolymerization will result in copolymers whose crystalline melting points vary monotonically with composition between those of the pure homopolymers. More common, however, is the case where the homopolymers form different crystal lattices because of steric differences. The random copolymerization of minor amounts of monomer B with A will disrupt the A lattice, lowering the T_m beneath that of homopolymer A, and vice versa. In an intermediate composition range the disruption will be so great that no crystallites can form, and the copolymers will be completely amorphous. A "phase diagram" for such a random copolymer is shown in Fig. 8.6.

The physical properties of random copolymers are determined by the region of the phase diagram, that is, by their composition and use temperature. In region 1, the polymer is a homogeneous, amorphous, and if pure, transparent material. The distinction between melt and rubbery behavior is not sharp — at higher temperatures the material flows more easily and becomes less elastic in character. It should be kept in mind, though, that the viscosities of polymer melts, even well above T_m or T_g, are far greater than those encountered in non-polymeric materials. A typical value might be on the order of a million centipoise.

A copolymer in region 5 is a typical amorphous, glassy polymer, hard, rigid, and usually brittle. Again, if the polymer is pure, it will be perfectly transparent. Polymethyl methacrylate (Lucite, Plexiglas) and polystyrene are familiar examples of homopolymers with these properties.

Copolymers in regions 2 and 3 consist of rigid crystallites dispersed in a relatively soft, rubbery, amorphous matrix. Since the refractive indices of the crystalline and amorphous phases differ, materials in these regions will be translucent to opaque, depending on the size of the crystallites, the degree of crystallinity, and the thickness of the sample. Since the crystallites restrict chain mobility, the materials are not elastic, but the rubbery matrix confers flexibility and toughness. The stiffness depends largely on the degree of crystallinity: the more rigid crystalline phase present, the stiffer the polymer. Polyethylene (squeeze bottles, bleach bottles, etc.) is a good example of a homopolymer with these properties, that is, one that is *between* its T_g and T_m at room temperature.

Copolymers in regions 4 and 6 consist of crystallites in an amorphous, glassy matrix. Since both phases are rigid, the materials are hard, stiff and rigid. Again, the two phases impart opacity. Nylons 6-6 and 6 are examples of homopolymers in a region below both T_g and T_m at room temperature.

8.8 GENERAL OBSERVATIONS ABOUT T_g AND T_m

Some other useful observations regarding T_m and T_g are that for polymers with a symmetrical repeating unit, for example, polyethylene $-\left[\begin{array}{cc} \overset{\text{H}}{\underset{\text{H}}{\text{C}}} & \overset{\text{H}}{\underset{\text{H}}{\text{C}}} \end{array}\right]_x -$ and polyvinylidene chloride (Saran) $-\left[\begin{array}{cc} \overset{\text{H}}{\underset{\text{H}}{\text{C}}} & \overset{\text{Cl}}{\underset{\text{Cl}}{\text{C}}} \end{array}\right]_x -$, $T_g/T_m \cong \frac{1}{2}$ and for unsymmetrical repeating units, for example, polypropylene $-\left[\begin{array}{cc} \overset{\text{H}}{\underset{\text{H}}{\text{C}}} & \overset{\text{CH}_3}{\underset{\text{H}}{\text{C}}} \end{array}\right]_x -$ and polychlorotrifluoroethylene (Kel-F) $-\left[\begin{array}{cc} \overset{\text{F}}{\underset{\text{F}}{\text{C}}} & \overset{\text{F}}{\underset{\text{Cl}}{\text{C}}} \end{array}\right]_x -$, $T_g/T_m \cong \frac{2}{3}$. In all cases $T_g < T_m$.

8.9 EFFECTS OF CROSSLINKING

To this point, the discussion has centered on noncrosslinked polymers. Light crosslinking, as in rubber bands, won't alter things appreciably. Higher degrees of crosslinking, however, if formed in the amorphous melt state, as is usually the case, will prevent the alignment of chains in a crystal lattice and hinder or prevent crystallization. Similarly, crosslinking restricts chain mobility and causes an increase in the apparent T_g. When the crosslinks are more frequent than every

40–50 main-chain atoms, the type of motion necessary to reach the rubbery state can never be achieved, and the polymer will degrade before reaching T_g.

8.10   OTHER TRANSITIONS

Transitions other than T_g and T_m sometimes are found in polymers. Some polymers possess more than one crystal form; so there will be an equilibrium temperature of transition from one to another. Similarly, second-order transitions below T_g have been observed in some materials (T_g is then termed the α transition, the next lower the β, etc.). These are attributed to motions of groups of atoms smaller than those necessary to produce T_g (Section 8.2, type 3 motions). These transitions may strongly influence properties. For example, *tough* amorphous plastics (e.g., polycarbonate) have such a transition well below room temperature, while *brittle* amorphous plastics (e.g., polystyrene and polymethyl methacrylate) do not.

REFERENCES

1.   P. Meares, *Trans. Faraday Soc.,* **53,** 31 (1957).
2.   L. Mandelkern, *Rubber Chem. Technol.,* **32,** 1392 (1959).
3.   A. V. Tobolsky, *Properties and Structure of Polymers,* Wiley, New York, 1960, Chapter 2.

PART 2

Polymer Synthesis

CHAPTER 9

Polycondensation
Reactions

9.1 INTRODUCTION

Polymerization reactions may be written generally as

$$x\text{-mer} + y\text{-mer} \rightarrow (x + y)\text{-mer} \tag{9.1}$$

In reactions that lead to what are normally considered condensation polymers, x and y may assume any value; that is, chains of any size may react together as long as they are capped with the complementary functional groups. These reactions are sometimes known as *step-growth* polymerizations. In addition polymerization, although x may assume any value, y is confined to unity. A growing chain can react only with a monomer molecule and continue its growth. Addition reactions are often termed *chain-growth* polymerizations. This is probably the most fundamental distinction between the two types of polymerization because it avoids the difficulty caused by the formation of condensation polymers without the elimination of a small molecule, for example, by ring scission, in which the small molecule has been eliminated previously in the formation of the cyclic monomer.

Regardless of the type of polymerization reaction, quantitative treatments are usually based on a common assumption, namely, that the reactivity of the functional group at a chain end is independent of the length of the chain. For example, in nylon 6-6,

$$H \left[\overset{H}{\underset{|}{N}} \text{-(CH}_2\text{)}_6\text{-N-}\overset{\overset{O}{\parallel}}{C}\text{-(CH}_2\text{)}_4 \; \overset{\overset{O}{\parallel}}{C} \right]_x OH$$

the rate constant for the reaction of the terminal amine and acid groups does not depend on x. Experimentally, this is an excellent assumption for x's greater than

103

about five or six. Since most polymers must develop x's on the order of a hundred or so to be of practical value, it introduces little error.

With these basic concepts in mind, we proceed to a more detailed and quantitative treatment of polymerization reactions.

9.2   STATISTICS OF POLYCONDENSATION

Consider the two equivalent linear polycondensation reactions, assuming difunctional monomers and stoichiometric equivalence.

$$\frac{x}{2}(\text{ARA}) + \frac{x}{2}(\text{BR'B}) \longrightarrow \text{ARA--[--BR'B--ARA--]}_{(x/2)-1}\text{BR'B*} \quad (9.2a)$$

$$x(\text{ARB}) \qquad \longrightarrow \text{ARB--[--ARB--]}_{x-2}\text{ARB} \qquad (9.2b)$$

It is important to note that x is used here to denote the *number of monomer residues* or *structural units* in the chain, rather than repeating units.

Each of the polymer molecules above contains a total of x A groups: $x - 1$ *reacted* A groups and 1 *unreacted* A group (on the end).

Let p = probability of finding a *reacted* A group, that is, the *conversion* or *extent of reaction*

$(1-p)$ = probability of finding an *unreacted* A group

N = total number of molecules present in the reaction mass (*of all sizes*)

n_x = number of molecules containing x A groups, both reacted and unreacted $\left(N = \sum\limits_{x=1}^{\infty} n_x\right)$

Now the total probability of finding a molecule with x A groups (reacted and unreacted) is equal to the mole or number fraction of those molecules present in the reaction mass n_x/N. This, in turn, is equal to the probability of finding a molecule with $(x - 1)$ reacted A groups and one unreacted A group, and since the total probability is the product of the individual probabilities,

$$\frac{n_x}{N} = p^{(x-1)}(1-p) = \text{mole (number) fraction } x\text{-mer} \qquad (9.3)$$

This is the so-called most probable distribution. It results from the random nature of the reaction between chains of different length. The distribution is plotted in Fig. 9.1 for several conversions. Note that the shorter chains are always more numerous, that is, the longer the chain length, the fewer the chains.

---

*Some of the molecules formed in this type of reaction will be capped with two A groups. For each of these, however, there will be one capped by two B groups; so (9.2a) represents the *average* reaction.

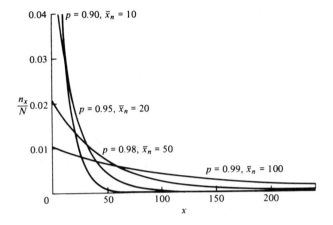

Figure 9.1   Number (mole) fraction distributions for linear polycondensations (9.3).

9.3   NUMBER-AVERAGE CHAIN LENGTHS

Since a distribution of chain lengths always arises in these condensation reactions, the average chain length is often of interest. Each molecule in the reaction mass has, on the average, one unreacted A group. If N_0 is the original number of molecules present in the reaction mass, at the start of the reaction there are N_0 unreacted A groups present, and at some time during the reaction there are N unreacted A groups; so $N_0 - N$ of the A groups have reacted. Thus

$$p = \frac{N_0 - N}{N_0} = \text{fraction of reacted A groups present} \qquad (9.4)$$

Because the N_0 original monomer molecules are distributed among the N molecules present in the reaction mass, the average number of monomer residues per chain, $\bar{x}_n$, is given by

$$\bar{x}_n = \frac{N_0}{N} = \text{average number of monomer residues per chain} \qquad (9.5)$$

Combining (9.4) and (9.5)

$$\bar{x}_n = \frac{1}{(1-p)} \qquad \text{(Carothers' equation)} \qquad (9.6)$$

This rather simple conclusion was reached by W. H. Carothers, the discoverer of nylon and one of the founders of polymer science, in the 1930s, but it is often ignored to this day. Its importance becomes obvious when it is realized that typical linear polymers must have $\bar{x}_n$'s on the order of 100 to achieve useful mechanical properties. This requires a conversion of *at least* 99%, assuming difunctional monomers in perfect stoichiometric equivalence. Such high conversions are almost unheard of in most organic reactions but are necessary in most cases to achieve high-molecular-weight condensation polymers. Since all reactions are reversible, many polycondensations would reach equilibrium at low conversions if the molecule of condensation were not efficiently removed (e.g., with heat and vacuum) to drive the reaction to high conversions.

Example 1.   Crud Chemicals is producing a linear polyester in a batch reactor by the condensation of a hydroxy acid HO—R—$\overset{\overset{\text{O}}{\|}}{\text{C}}$—OH. It has been proposed to follow the progress of the reaction by measuring the amount of water removed from the reaction mass. Assuming pure monomer and that all the water condensation can be removed and measured, derive for them an equation relating $\bar{x}_n$ of the polymer to N_0, the moles of monomer charged, and M, the total moles of water evolved since the start of the reaction.

Solution.   In this polycondensation reaction, each reacted —OH (or —$\overset{\overset{\text{O}}{\|}}{\text{C}}$—OH) group produces 1 molecule of water. Therefore, the moles of reacted —OH (or —$\overset{\overset{\text{O}}{\|}}{\text{C}}$—OH) groups is M. The probability of finding a reacted group (of either kind) is

$$p = \frac{M}{N_0} = \frac{\text{moles reacted —OH groups}}{\text{total moles —OH groups}}$$

Plugging into Carothers' equation (9.6)

$$\bar{x}_n = \frac{1}{1 - (M/N_0)}$$

If the restriction on perfect stoichiometric equivalence is removed, similar (but more involved) reasoning leads to

$$\bar{x}_n = \frac{1 + r}{2r(1 - p) + (1 - r)} = \frac{1 + r}{1 + r - 2rp} \tag{9.7}$$

where r is the stoichiometric ratio of functional groups present, N_{A_0}/N_{B_0} (A is taken to be the limiting reagent, so r is always less than 1, and p represents the fraction of A groups reacted).

To examine the effect of stoichiometric imbalance, consider the limiting case of complete conversion, $p = 1$, for which (9.7) reduces to

$$\bar{x}_n = \frac{1 + r}{1 - r} \quad (\text{for } p = 1) \tag{9.8}$$

At $r = 1$, all the reactant molecules are combined in a single molecule of essentially infinite molecular weight. Reducing r to 0.99 cuts $\bar{x}_n$ to 199; reducing r to 0.95 cuts $\bar{x}_n$ all the way down to 39. So to reach high chain lengths, a close approach to stoichiometric equivalence is required in addition to high conversions. Considering the usual industrial purity levels and the precision of weighing techniques, this is not always easy to achieve. Fortunately, in the production of nylons from diacids and diamines, the monomers form an ionic salt of perfect 1–1 ratio which is carefully purified before polymerization. Also, chain length may sometimes be pushed up despite a stoichiometric imbalance by an *ester interchange* reaction. For example, if a polyester is formed from a diacid plus an excess of glycol, the reaction will stop at a point when all the chains are capped by —OH's. If the glycol is volatile enough to be driven off with the application of heat and vacuum, the chains may combine further through ester interchange with the elimination of excess glycol.

$$H\text{-}O\text{-}R\text{-}O\left[\!\!\begin{array}{c}O \quad O \\ \| \quad \| \\ C\text{-}R'\text{-}C\text{-}O\text{-}R\text{-}O\end{array}\!\!\right]_x \!\!H + H\!\!\left[\!\!\begin{array}{c}O \quad O \\ \| \quad \| \\ O\text{-}R\text{-}O\text{-}C\text{-}R'\,C\end{array}\!\!\right]_y \!\!O\text{-}R\text{-}OH \longrightarrow$$

$$HOR\text{-}O\left[\!\!\begin{array}{c}O \quad O \\ \| \quad \| \\ C\text{-}R'\text{-}C\text{-}O\text{-}R\text{-}O\end{array}\!\!\right]_{(x+y)} \!\!H \quad + HOROH$$

On the other hand, introduction of an excess of one of the monomers or of some monofunctional material to reduce r deliberately provides a convenient means of limiting chain length.

9.4   CHAIN LENGTHS ON A WEIGHT BASIS

The previous development of the distribution of chain lengths in a linear condensation polymer, although perfectly legitimate, is in some ways misleading because it describes the *number* of molecules of a given chain length present and counts equally both monomer units and chains containing many hundreds of monomer units; that is, each is one molecule. For example, in a mixture consisting of one monomer molecule and one 100-mer, the *number* or *mole* fraction is 1/2. Another way of looking at it is to inquire about the relative *weights* of the various chain lengths present. On this basis, the *weight fraction* monomer in the mixture is only 1/101. By neglecting the weight of the small molecule split out in most condensation reactions, we can obtain the distribution of chain lengths in terms of weight fractions. If $(w_x/W) =$ weight fraction of x-mer present = weight of x-mers/weight of batch, letting $m =$ average molecular weight of a structural unit,

$$\text{weight of } x\text{-mer} = mxn_x \tag{9.9}$$

$$\text{weight of batch} = mN_0 \tag{9.10}$$

$$\frac{w_x}{W} = \frac{xn_x}{N_0} \tag{9.11}$$

Combining (9.3), (9.4), and (9.11) and eliminating the N's gives

$$\frac{w_x}{W} = xp^{(x-1)}(1-p)^2 \tag{9.12}$$

This "most probable" *weight fraction* distribution is shown in Fig. 9.2. Although the monomer molecules are the most numerous, their combined weight is an insignificant portion of the total weight. The peak is at $x = -1/\ln p \cong 1/(1-p)$; that is, the x-mer present in greatest weight is *approximately* $\bar{x}_n$. The neglect of the molecule of condensation in this derivation can lead to significant errors at low x's. The exact solution is quite complex.[1]

The weight-average chain length, $\bar{x}_w$ can be obtained by inserting the number distribution (9.3) into (6.25) and integrating at constant p.

$$\bar{x}_w = \frac{\int_0^\infty (n_x/N)x^2\,dx}{\int_0^\infty (n_x/N)x\,dx} = \frac{\int_0^\infty x^2 p^{(x-1)}\,dx}{\int_0^\infty xp^{(x-1)}\,dx} = \frac{1+p}{1-p} \tag{9.13}$$

(Evaluation of the above integrals is a good exercise for mathematical masochists.) Combining (9.13) and (9.6) gives an expression for the polydispersity index in

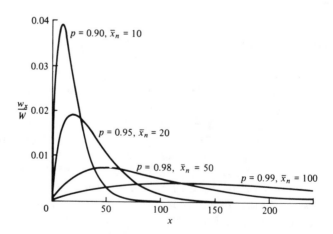

Figure 9.2   Weight fraction distributions for linear polycondensations.[9,12]

terms of conversion

$$\frac{\bar{x}_w}{\bar{x}_n} = (1 + p) \qquad (0 \leqslant p \leqslant 1) \tag{9.14}$$

9.5 GEL FORMATION[2]

If a condensation polymerization batch contains some monomer with a functionality $f > 2$, and if the reaction is carried to a high enough conversion, a cross-linked network or *gel* may be formed. For the purposes of this discussion, a gel is defined as a molecule of essentially infinite molecular weight, extending throughout the reaction mass. In the production of thermosetting condensation polymers, the reaction must be terminated short of the conversion at which a gel is formed, or the product could not be molded or processed further (cross-linking is later completed in the mold). Hence, the prediction of this *gel-point* conversion is of great practical importance.

The multifunctional monomer is represented by A_f, for example,

$$f = 3 \qquad\qquad f = 4$$

In a gel network the multifunctional monomer acts as a *branch unit.* These branch units are connected to *chains,* that is, linear segments of difunctional units that lead *either* to another branch unit *or* to an unreacted end:

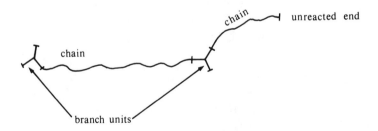

The *branching coefficient* α is defined as the probability that a given functional group on a branch unit is connected to another branch unit by a chain (rather than to an unreacted end).

Further analysis is based on two assumptions: (a) All functional groups are equally reactive. This might not always be true. For example, the secondary

(middle) hydroxyl group in glycerine, HO—C—C—C—OH, is probably not as

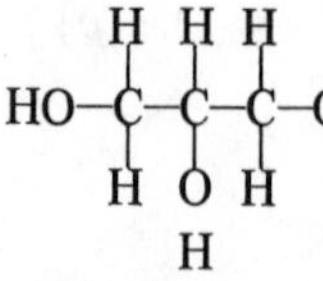

reactive as the primary (end) hydroxyls. (b) Furthermore, intramolecular condensations (i.e., reactions between A and B groups on the same molecule) do not occur. In other words, each reaction between an A and a B reduces the number of molecules in the reaction mass by one.

Consider the reaction

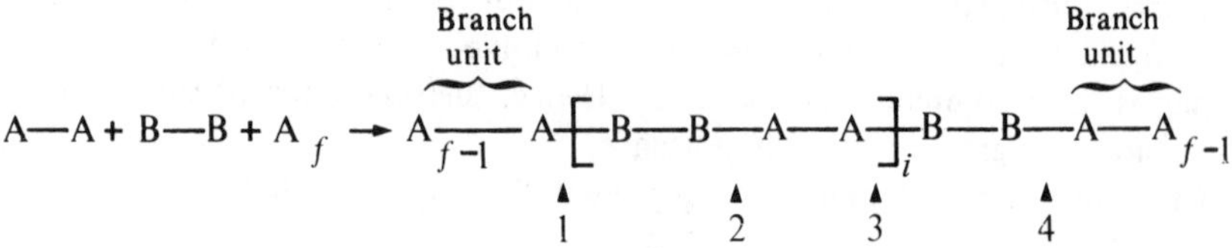

$$A\!-\!A + B\!-\!B + A_f \longrightarrow A\frac{}{f-1}A\!\left[\!B\!-\!B\!-\!A\!-\!A\!\right]_i\!B\!-\!B\!-\!A\!-\!A_{f-1}$$

We define ρ = fraction of the *total* A groups in the branch units A_f

$(1 - \rho)$ = fraction of the total A groups in the difunctional A—A molecules

p_A = probability of finding a reacted A group

p_B = probability of finding a reacted B group

The probabilities of finding each of the numbered types of bonds in the molecule shown on the right above are tabulated as follows:

Bond	Probability of formation
1	p_A, probability of A reacting with B
2	$p_B(1 - \rho)$, probability of B reacting with A on A—A
3	p_A, probability of A reacting with B
4	$p_B \rho$, probability of B reacting with branch unit A

Therefore the overall probability of finding the chain shown is

$$p_A[p_B(1 - \rho)p_A]^i p_B \rho$$

Since i may vary from $0 \to \infty$, the probability of finding one of *all* such chains connecting branch units is

$$\alpha = \sum_{i=0}^{\infty} p_A[p_B(1 - \rho)p_A]^i p_B \rho = \frac{p_A p_B \rho}{1 - p_A p_B(1 - \rho)} \qquad (9.15)$$

Now

$$r = \frac{N_{A_0}}{N_{B_0}}, \tag{9.16}$$

$$p_A = \frac{N_{A_0} - N_A}{N_{A_0}}, \tag{9.17}$$

$$p_B = \frac{N_{B_0} - N_B}{N_{B_0}} \tag{9.18}$$

From the stoichiometry of the reaction

$$N_{A_0} - N_A = N_{B_0} - N_B \tag{9.19}$$

therefore

$$p_B = \frac{N_{A_0}}{N_{B_0}} p_A = r p_A \tag{9.20}$$

and

$$\alpha = \frac{r p_A^2 \rho}{1 - r p_A^2 (1 - \rho)} = \frac{p_B^2 \rho}{r - p_B^2 (1 - \rho)} \tag{9.21}$$

Each of the terminal branch units in the molecule on the right above has $(f-1)$ unreacted functional groups. If *at least* one of these unreacted A groups is then connected up to another branch unit, a gel is formed. Since all the unreacted A groups on branch units are statistically identical, *all* the branch units in the reaction mass will be connected together by chains. This occurs when $\alpha(f-1) \geqslant 1$, that is, when the product of the probability of a chain connecting a given functional group to another branch unit and the number of remaining unreacted functional groups becomes a certainty. Thus the critical value of α for gelation is

$$\alpha_c = \frac{1}{f-1} \tag{9.22}$$

Combining (9.22) and (9.21) gives the gel-point conversion.

Example 2. Calculate the gel-point conversion of a reaction mass consisting of 2 moles of glycerine plus 3 moles of phthalic anhydride:

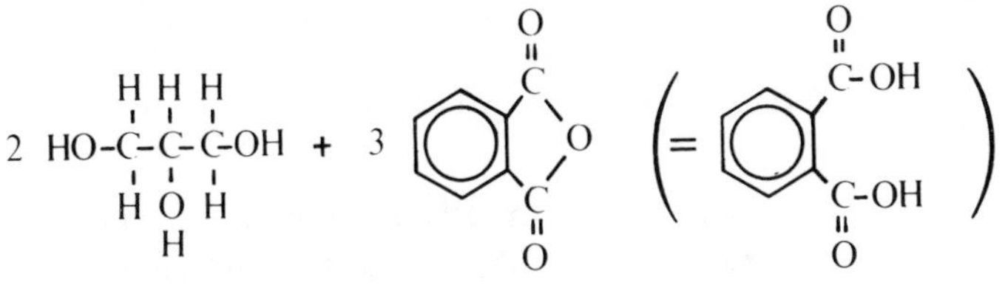

Solution. By convention, A should be chosen as the functional group on the multifunctional monomer. Here, $A = OH$, $f = 3$ and $\alpha_c = 0.5$ (9.22). Also, $N_{A_0} = N_{B_0}$, so $r = 1$. Since *all* the A ($-OH$) groups are on the multifunctional (glycerine) molecules, $\rho = 1$. Plugging these values into (9.21) and solving gives $p_A = 0.707$.

Experimentally, the gel-point conversion for this system was found to be $p_A = 0.765$. This was done by noting the conversion at which bubbles ceased to rise in the reaction mass. The difference between the experimental and theoretical results can be accounted for by some intramolecular condensation and the lower reactivity of the secondary hydroxyls. Also, Bobalek et al.[3] showed that as large molecules grew within the reaction mass, their molecular weights became high enough to cause them to precipitate from solution before a true infinite gel network was formed (a particle just 1 μm in diameter has a molecular weight of the order of 10^{11} g/g mole). Higher conversions were needed to link these precipitated particles together to prevent the rise of bubbles.

Regardless of the reasons, however, the theoretical calculation usually gives the *lower limit* or most conservative value for the gel-point conversion, providing a margin of safety in practice. An apparent exception sometimes occurs in the preparation of unsaturated polyester resins (see Chapter 2, Example 3). In these calculations, the double bond would be assumed inert in the condensation reaction. Under certain conditions some addition crosslinking through the double bonds occurs inadvertently during the condensation reaction. This, in effect, makes f much greater than that for condensation alone, which results in premature gelling. It is normally counteracted by incorporating an inhibitor (see Chapter 10) for addition reactions during the polycondensation.

Example 3. Calculate the *minimum* number of moles of ethylene glycol (I) needed to produce a gel when reacted with one mole of phthalic anhydride (II) and one mole of BTDA (III)

$$HO-CH_2-CH_2-OH$$

(I)   (II)   (III)

Solution. First, recognize that BTDA (III) is a *di*anhydride and therefore can be considered a tetrafunctional acid. Again, A is by convention the functional group on the multifunctional monomer, here $A = -\overset{\overset{\displaystyle O}{\parallel}}{C}-OH$, and $(I) = B-B$, $(II) = A-A$, and $(III) = A_4$. The tendency toward gelation increases with conversion, and with the minimum moles of (I) present, A groups will be in stoichiometric excess — that is, B is the limiting reagent. Therefore

$$(p_B)_{max} = 1.0 \text{ for minimum (I)}$$

$$\alpha_c = \frac{1}{f-1} = \frac{1}{4-1} = \frac{1}{3} \tag{9.22}$$

Here,

$$\rho = \frac{\text{moles A on (III)}}{\text{moles A on (II)} + \text{(III)}} = \frac{4}{6} = \frac{2}{3}$$

At the gel point, with $p_B = 1$ and $\rho = 2/3$

$$\frac{1}{3} = \frac{p_B^2 \rho}{r - p_B^2(1-\rho)} = \frac{2/3}{r - (1/3)} \tag{9.21}$$

$$r = \frac{7}{3} = \frac{N_{A_0}}{N_{B_0}}$$

$$N_{B_0} = \frac{N_{A_0}}{r} = 6 \times \frac{3}{7} = \frac{18}{7}$$

but $N_{B_0} = 2(N_I)$; that is, there are 2 moles of hydroxyl per mole of glycol:

$$N_I = \frac{N_{B_0}}{2} = \frac{18}{14} = \frac{9}{7} \text{ moles}$$

9.6   KINETICS OF POLYCONDENSATION

The kinetics of polycondensation reactions is similar to that of ordinary condensation reactions. Since the average chain length is related to conversion in linear polycondensation by (9.7), and conversion is given as a function of time by the kinetic expression, $\bar{x}_n$ is directly related to the reaction time and can thus be controlled by limiting the reaction time. Similarly, the time to reach a gel point is related by the rate expressions (9.23) and (9.24).

Example 4.   For a second-order irreversible polycondensation reaction, with rate proportional to the concentrations of reactive A and B groups, obtain expressions for conversion and number-average chain length as a function of time for a stoichiometrically equivalent batch.

Solution.   For a second-order reaction, assuming constant volume,

$$\frac{-d[A]}{dt} = k[A][B], \tag{9.23}$$

where k is the reaction-rate constant and the brackets indicate concentrations. For $[A_0] = [B_0]$, that is, equivalent initial concentration, $[A] = [B]$ at all times, and

$$\frac{-d[A]}{dt} = k[A]^2 \tag{9.24}$$

Separating variables and integrating between the limits when $t = 0$, $[A] = [A_0]$, and when $t = t$, $[A] = [A]$, gives

$$\frac{1}{[A]} - \frac{1}{[A_0]} = kt \tag{9.25}$$

since

$$p = \frac{[A_0] - [A]}{[A_0]} = \frac{[A_0]kt}{1 + [A_0]kt} \tag{9.26}$$

Combining (9.6) and (9.26),

$$\bar{x}_n = 1 + [A_0]kt \tag{9.27}$$

That is, the number-average chain length increases linearly with time.

Two cautions are in order about the preceding example. First, by writing an *irreversible* rate expression, we are assuming that any molecule of condensation is being continuously and efficiently removed from the reaction mass. Second, not all polycondensations are second order. Some polyesterifications, for example, are catalyzed by their own acid groups and are therefore first order in hydroxyl concentration, second order in acid, and third order overall. The rate may also be proportional to the concentration of an added catalyst (usually acids or bases), if used.

REFERENCES

1.   H. E. Grethlein, *Ind. Eng. Chem. Fundame.* 8, 206, 1969.
2.   P. J. Flory, *Principles of Polymer Chemistry,* Cornell University Press, Ithaca, N.Y., 1953.
3.   E. G. Bobalek et al., *J. Appl. Polym. Sci.,* 8, 625 (1964).

CHAPTER 10

Free-radical Addition Polymerization

10.1 INTRODUCTION

One of the most important types of addition polymerization is initiated by the action of *free radicals*, electrically neutral species with an *unshared electron*. In the developments to follow, a dot · will represent a single electron. The single bond, a pair of shared electrons, will be denoted by a double dot :, or where not necessary to indicate electronic configurations, by the usual −. A double bond, *two* shared electron pairs, is : : or =.

Free radicals for the initiation of addition polymerization are usually generated by the thermal decomposition of organic peroxides or azo compounds. Two common examples are

Benzoyl peroxide

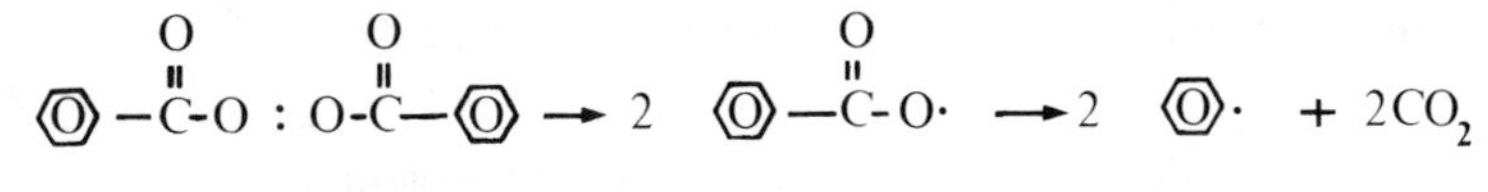

$$\underset{\bigcirc}{}-\overset{\overset{O}{\|}}{C}\text{-O : O-}\overset{\overset{O}{\|}}{C}-\bigcirc \longrightarrow 2\ \bigcirc-\overset{\overset{O}{\|}}{C}\text{-O·} \longrightarrow 2\ \bigcirc\cdot\ +\ 2CO_2$$

Azobisisobutyronitrile

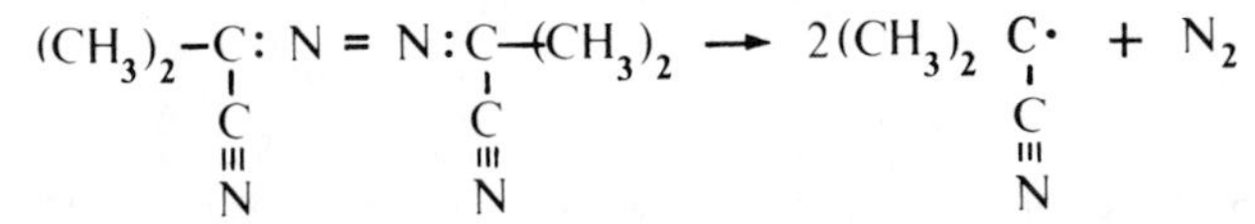

$$(CH_3)_2-\underset{\underset{N}{\overset{\|}{\underset{N}{C}}}}{C}: N = N : \underset{\underset{N}{\overset{\|}{\underset{N}{C}}}}{C}-(CH_3)_2 \longrightarrow 2(CH_3)_2\ \underset{\underset{N}{\overset{\|}{\underset{N}{C}}}}{C}\cdot\ +\ N_2$$

10.2   MECHANISM OF POLYMERIZATION

The initiator molecule, represented by I, decomposes by a first-order reaction with a rate constant k_d to give two free radicals $R\cdot$, as follows:

$$I \xrightarrow{k_d} 2R\cdot \qquad \text{(decomposition)} \qquad (10.1)$$

The radical then adds a monomer by grabbing an electron from the electron-rich double bond, forming a single bond with the monomer but leaving an unshared electron at the other end:

$$R\cdot \ + \ \begin{matrix} H & H \\ | & | \\ C & : : C \\ | & | \\ H & X \end{matrix} \ \longrightarrow \ R:\begin{matrix} H & H \\ | & | \\ C & : C\cdot \\ | & | \\ H & X \end{matrix}$$

This may be abbreviated by

$$R\cdot + M \xrightarrow{k_a} P_1\cdot \qquad \text{(addition)} \qquad (10.2)$$

where $P_1\cdot$ represents a growing polymer radical with one repeating unit. Note that the product of the addition reaction is still a free radical; it proceeds to propagate the chain by adding another monomer unit

$$P_1\cdot + M \xrightarrow{k_p} P_2\cdot$$

again maintaining the unshared electron at the chain end, which adds another monomer unit

$$P_2\cdot + M \xrightarrow{k_p} P_3\cdot$$

and so on. In general, the propagation reaction is written as

$$P_x\cdot + M \xrightarrow{k_p} P\cdot_{(x+1)} \qquad \text{(propagation)} \qquad (10.3)$$

We have again assumed that reactivity is independent of chain length by using the same k_p for each propagation step.

A growing chain can be terminated in one of two ways:

$$\sim\sim\begin{matrix} H & H \\ | & | \\ C & -C\cdot \\ | & | \\ H & X \end{matrix} \ + \ \begin{matrix} H & H \\ | & | \\ \cdot C & -C \\ | & | \\ X & H \end{matrix}\sim\sim \ \longrightarrow \ \sim\sim\begin{matrix} H & H & H & H \\ | & | & | & | \\ C & -C & :C & -C \\ | & | & | & | \\ H & X & X & H \end{matrix}\sim\sim$$

$$P_x \cdot + P_y \cdot \xrightarrow{k_{tc}} P_{(x+y)} \qquad (\text{combination}) \qquad (10.4a)$$

where $P_{(x+y)}$ is a *dead* polymer chain of $(x+y)$ repeating units.

$$P_x \cdot + P_y \cdot \xrightarrow{k_{td}} P_x + P_y \qquad (\text{disproportionation}) \qquad (10.4b)$$

The relative proportion of each termination mode depends on the particular polymer and the reaction temperature, but in most cases one or the other predominates.

Example 1. A free radical initiator $R:R(R:R \rightarrow 2R \cdot)$ and a vinyl monomer

H H
| |
C=C are polymerized in a homogeneous reaction mass. The resulting molecules
| |
H X

all have the structures

(I)   (II)

Tabulate all possible combinations of (a) groups A and B and (b) of groups D and E. Interchanging A and B or D and E are not considered separate combinations.

Solution. Molecules of type I are terminated by disproportionation; those of type II by combination. Therefore

(a).

A	B
R	H
R	$-\overset{\displaystyle H}{\underset{\displaystyle X}{C}}=\overset{\displaystyle H}{C}$

(b).

D	E
R	R

10.3   KINETICS OF HOMOGENEOUS POLYMERIZATION

In practice, not all the radicals generated in reaction (10.1) actually initiate chain growth as in reaction (10.2). Some recombine or are used up by side reactions. Letting f be the fraction of radicals generated that actually do initiate chain growth, the rate of formation of growing chain radicals through reactions (10.1) and (10.2) is

$$r_i = \left(\frac{d\,[P_1 \cdot]}{dt}\right)_i = 2fk_d\,[I] \qquad \text{(rate of initiation)} \qquad (10.5)$$

Although it is rarely mentioned explicitly, this expression is based on the generally valid assumption that the rate of reaction (10.2) is much greater than that of (10.1), that is, that the initiator decomposition is *rate controlling*. Thus, as soon as an initiator radical is formed, it grabs a monomer molecule, starting chain growth; so k_a does not appear in the expression.

According to reaction (10.3), the rate of monomer removal in the propagation step is

$$r_p = -\left(\frac{d\,[M]}{dt}\right)_p = k_p\,[M]\,[P\cdot] \qquad \text{(rate of propagation)} \qquad (10.6)$$

where $[P\cdot]$ is the *total* concentration of growing chain radicals; that is,

$$[P\cdot] = \sum_{x=1}^{\infty}\,[P_x\cdot] \qquad (10.7)$$

The rate of removal of chain radicals is the sum of the rates of the two termination reactions. Since both are second order

$$r_t = -\frac{d\,[P\cdot]}{dt} = 2k_t[P\cdot]^2 \qquad \text{(rate of termination)} \qquad (10.8)$$

where

$$k_t = (k_{tc} + k_{td}) \qquad (10.9)$$

Unfortunately, the unknown quantity $[P\cdot]$ is present in the equations. It can be removed by making the standard kinetic assumption of a steady-state concentration of a transient species, in this case, the chain radicals. In order that $[P\cdot]$ remain constant, chain radicals must be generated at the same rate at which they are removed, $r_i = r_t$, or

$$2fk_d\,[I] = 2k_t[P\cdot]^2 \qquad (10.10)$$

This gives the chain-radical concentration

$$[P \cdot] = \left(\frac{fk_d[I]}{k_t} \right)^{1/2} \tag{10.11}$$

which, when inserted in (10.6), provides an expression for the rate of monomer removal in the propagation reaction

$$r_p = -\left(\frac{d[M]}{dt} \right)_p = k_p \left(\frac{fk_d[I]}{k_t} \right)^{1/2} [M] \tag{10.12}$$

Actually, monomer is consumed in the addition step (10.2) also, but for long chains the amount is insignificant, so that (10.12) may be taken as the overall rate of polymerization — the rate at which monomer is converted to polymer.

Equation (10.12) is the classical rate expression for a *homogeneous* free-radical polymerization. It has been well substantiated in many important cases, particularly in fairly dilute solutions. In certain cases of industrial importance, however, significant deviations are observed. These are discussed in Chapter 13 on polymerization practice.

Integration of (10.12) gives the conversion $X = ([M_0] - [M])/[M_0]$ as a function of time. A form occasionally used in studying the initial stages of an isothermal batch reaction is obtained by assuming constant initiator concentration $[I_0]$.

$$\ln \frac{[M]}{[M_0]} = -k_p \left(\frac{fk_d[I_0]}{k_t} \right)^{1/2} t \tag{10.13}$$

The assumption of constant $[I_0]$ may not be realistic in cases of industrial importance at high conversions, however. By including the first-order decay of the initiator, starting at time $t = 0$, with $[I] = [I_0]$,

$$\frac{d[I]}{dt} = -k_d[I] \tag{10.14a}$$

$$[I] = [I_0]e^{-k_d t} \tag{10.14b}$$

and inserting (10.14b) into (10.12) and integrating, a more realistic expression for monomer concentration versus time is obtained:[*]

---

[*] All integrations here have been performed assuming a constant-volume reaction mass. This is not strictly true, as most addition polymerization systems undergo a density increase of 10–20% in going from a liquid monomer to polymer (carrying out the reaction in an inert solvent reduces this, of course). The effect of volume change may be incorporated by assuming that the volume is linear in conversion, as outlined by Levenspeil (1). For the first-order reactions leading to (10.13) and (10.14b), the result is the same, but even this simple assumption essentially precludes an analytic counterpart of (10.15).

$$\ln \frac{[M]}{[M_0]} = \frac{2k_p}{k_d}\left(\frac{fk_d[I_0]}{k_t}\right)^{1/2}[e^{-(k_dt/2)} - 1] \tag{10.15}$$

This expression has some interesting and important implications when compared with (10.13). Setting $t = \infty$ reveals that there is a maximum attainable conversion that depends on $[I_0]$:

$$\text{Maximum conversion} = \left(1 - \frac{[M]}{[M_0]}\right)_{max} = 1 - \exp\left\{-\frac{2}{k_d}\left(\frac{k_p^2}{k_t}fk_d[I_0]\right)^{1/2}\right\} \tag{10.16}$$

This "dead-stop" situation is basically a matter of the initiator being used up before the monomer, but it is not revealed by (10.13), which always predicts complete conversion at long enough times. Thus the use of (10.13) instead of (10.15) can result in considerable error at high conversions, although they approach each other at low conversions.

Example 2.   For the polymerization of pure styrene with azobisisobutyronitrile, at 60°C, in an isothermal batch reactor,
a.   Compare (10.13) and (10.15) by plotting conversion $1 - [M]/[M_0]$ versus time, using the data given below.
b.   Determine the minimum $[I_0]$ needed to achieve 90% conversion. Notice that according to (10.13), there is *no* minimum.
Data: $k_p^2/k_t = 1.18 \times 10^{-3}$ liter/mole sec, ρ styrene $= 0.907\,\text{g/cm}^3$

$$k_d = 0.96 \times 10^{-5}\ \text{sec}^{-1}$$

$$[I_0] = 0.05\ \text{mole/liter}$$

$$f = 1.0$$

Solution
a.   The molecular weight of styrene is 104. Therefore,

$$[M_0] = 0.907\frac{\text{g}}{\text{cm}^3} \times \frac{1000\ \text{cm}^3}{\text{liter}} \times \frac{1\ \text{g mole}}{104\ \text{g}} = 8.7\frac{\text{g moles}}{\text{liter}}$$

Plugging this and the data into (10.13) and (10.15) results in the plots shown in Fig. 10.1. The maximum conversion under these conditions is 99.3% (10.16).
b.   Setting $[1 - ([M]/[M_0])]_{max} = 0.90$ in (10.16), and solving gives an $[I_0]_{min} = 0.0108$ mole/liter.

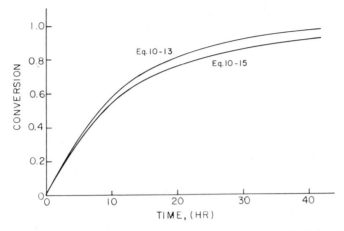

Figure 10.1 Conversion versus time for isothermal free-radical batch polymerization (data of Example 2).

Side reactions sometimes cause deviations from the classical kinetics. Agents that cause these reactions are generally categorized as *inhibitors* or *retarders*. An inhibitor delays the start of the reaction, but once begun, it proceeds at the normal rate. Vinyl monomers are normally shipped with a few parts per million inhibitor to prevent polymerization in transit. A retarder slows down the reaction rate. Some chemicals combine both effects. These are illustrated in Fig. 10.2. Oxygen is often an effective inhibitor for the free-radical polymerization of vinyl monomers; so the reactions are normally carried out under a blanket of nitrogen.

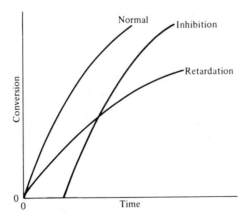

Figure 10.2   Inhibition and retardation.

10.4   INSTANTANEOUS AVERAGE CHAIN LENGTHS

As in condensation polymerization, a distribution of chain lengths is always obtained in a free-radical addition polymerization because of the inherently random nature of the termination reaction with regard to chain length. Expressions for the number-average chain length are usually couched in terms of the *kinetic chain length* ν, which is the rate of monomer addition to growing chains over the rate at which chains are started by initiator radicals, that is, the average number of monomer units per growing chain radical at a particular instant. It thus expresses the efficiency of the initiator radicals in polymerizing monomer.

$$\nu = \frac{r_p}{r_i} = \frac{k_p[M]}{2(fk_dk_t)^{1/2}[I]^{1/2}} \tag{10.17}$$

If the growing chains terminate exclusively by disproportionation, they undergo no change in length in the process, but if combination is the exclusive mode of termination, the growing chains, on the average, double in length upon termination. Therefore

$$\bar{x}_n = \nu \qquad \text{(termination by disproportionation)} \tag{10.18a}$$

$$\bar{x}_n = 2\nu \qquad \text{(termination by combination)} \tag{10.18b}$$

The average chain length may be expressed more generally in terms of a quantity ξ, the average number of dead chains produced per termination.

ξ is the rate of dead chain formation per rate of termination reactions. Since each disproportionation reaction produces *two* dead chains and each combination reaction *one*,

$$\text{rate of dead chain formation} = (2k_{td} + k_{tc})[P\cdot]^2 \tag{10.19}$$

$$\text{rate of termination reactions} = (k_{tc} + k_{td})[P\cdot]^2 \tag{10.20}$$

$$\xi = \frac{k_{tc} + 2k_{td}}{k_{tc} + k_{td}} = \frac{k_{tc} + 2k_{td}}{k_t} \tag{10.21}$$

The instantaneous number-average chain length is the rate of addition of monomer units to all chains (r_p) over the rate of dead chain formation:

$$\bar{x}_n = \frac{k_p[P\cdot][M]}{(2k_{td} + k_{tc})[P\cdot]^2} = \frac{k_p[M]}{\xi(fk_dk_t)^{1/2}[I]^{1/2}} \tag{10.22}$$

When $k_{td} \gg k_{tc}$, $\xi = 2$, and when $k_{tc} \gg k_{td}$, $\xi = 1$, duplicating the previous result; but (10.22) can also handle various degrees of mixed termination. This is particularly important when considering nonisothermal reactions where ξ may change appreciably with temperature.

Keeping in mind that $\bar{x}_n$ is one of the most important factors in determining certain mechanical properties of polymers, what is the significance of (10.22)? A growing chain may react with another growing chain and terminate, or it may add another monomer unit and continue its growth. The more monomer molecules in the vicinity of the chain radical, the higher the probability of another monomer addition; hence the proportionality to [M]. On the other hand, the more initiator radicals there are present competing for the available monomer, the shorter the chains will be, on the average, causing the inverse proportionality to the square root of [I]. Equation (10.12) shows that the rate of polymerization can be increased (always a desirable economic goal) by increasing both [M] and [I], the former being more efficient than the latter. However, there is an upper limit to [M] set by the density of the pure monomer at the reaction conditions. So it is often tempting to increase [I] to attain higher rates. But according to (10.22), this unavoidably lowers $\bar{x}_n$. You can't have your cake and eat it too.

10.5   TEMPERATURE DEPENDENCE OF RATE AND CHAIN LENGTH

We may assume that the temperature dependence of the individual rate constants in (10.12) is given by the Arrhenius expression

$$k_i = A_i e^{-E_i/RT} \tag{10.23}$$

where k_i is the rate constant for a particular elementary reaction, A_i its frequency factor, and E_i its activation energy. If we neglect the temperature dependence of f, (10.12) becomes

$$r_p = C \exp \left\{ - \left[\frac{E_p + (E_d/2) - (E_t/2)}{RT} \right] \right\} \tag{10.24}$$

where C combines temperature-independent quantities into a single constant. By taking the logarithm of r_p we get

$$\ln r_p = \ln C - \left\{ \left[\frac{E_p + (E_d/2) - (E_t/2)}{RT} \right] \right\} \tag{10.25}$$

and differentiating with respect to T gives

$$\frac{d \ln r_p}{dT} = \left[\frac{E_p + (E_d/2) - (E_t/2)}{RT^2} \right] \tag{10.26}$$

Similar treatment of (10.22), with the assumption that ξ is independent of temperature, results in

$$\frac{d \ln \bar{x}_n}{dT} = \left[\frac{E_p - (E_d/2) - (E_t/2)}{RT^2} \right] \qquad (10.27)$$

Now for a typical free-radical addition reaction, $[E_p - (E_t/2)] \cong 5 \, \text{kcal/g mole}$ and $E_d \cong 30 \, \text{kcal/g mole}$. Thus

$$\frac{d \ln r_p}{dT} \cong \frac{+20 \, \text{kcal/g mole}}{RT^2} \qquad (10.28)$$

and

$$\frac{d \ln \bar{x}_n}{dT} \cong \frac{-10 \, \text{kcal/g mole}}{RT^2} \qquad (10.29)$$

Example 3.   Estimate the changes in r_p and $\bar{x}_n$ for a typical homogeneous free-radical addition polymerization on going from 60°C to 70°C.

Solution.   Integration of (10.26) between T_1 and T_2 gives

$$\ln \frac{r_p(T_2)}{r_p(T_1)} = \frac{-[E_p + (E_d/2) - (E_t/2)]}{R} \left[\frac{1}{T_2} - \frac{1}{T_1} \right] \qquad (10.30)$$

With the approximate activation energies above, $T_1 = 273 + 60 = 333 \, \text{K}$, $T_2 = 273 + 70 = 343 \, \text{K}$, and $R = 1.99 \, \text{cal/g mole K}$

$$\ln \frac{r_p \, (70°C)}{r_p \, (60°C)} = 0.879 \qquad \text{and} \qquad \frac{r_p \, (70°C)}{r_p \, (60°C)} = 2.41 \, (!)$$

so the rate increases by 141% over the stated 10° temperature interval. This rather spectacular (and sometimes dangerous) sensitivity is due mainly to the large activation energy for the rate-controlling initiator decomposition. For $\bar{x}_n$

$$\ln \frac{\bar{x}_n \, (T_2)}{\bar{x}_n \, (T_1)} = \frac{-[E_p - (E_d/2) - (E_t/2)]}{R} \left[\frac{1}{T_2} - \frac{1}{T_1} \right] \qquad (10.31)$$

$$\ln \frac{\bar{x}_n \, (70°C)}{\bar{x}_n \, (60°C)} = -0.439 \qquad \text{and} \qquad \frac{\bar{x}_n \, (70°C)}{\bar{x}_n \, (60°C)} = 0.644$$

Thus $\bar{x}_n$ is subject to a 35.6% *decrease* over the same temperature range.

10.6   CHAIN TRANSFER

In practice, another type of reaction often occurs in free-radical addition polymerizations. These *chain-transfer* reactions kill a growing chain radical and start a new one in its place:

$$R':H + P_x \cdot \xrightarrow{k_{tr}} P_x + R' \cdot \qquad (10.32a)$$

$$R' \cdot + M \xrightarrow{k'_a} P_1 \cdot, \text{ etc.} \qquad (10.32b)$$

Thus chain transfer results in shorter chains, and if reactions (10.32) are not too frequent compared to the propagation reaction and don't have very low rate constants, it will not change the overall rate of polymerization appreciably.

The compound $R':H$ is known as a *chain-transfer agent*. Under appropriate conditions, almost anything in the reaction mass may act as a chain-transfer agent, including initiator, monomer, solvent, and dead polymer.

Example 4.   Show, using dots to represent the electrons involved, how chain transfer to dead polymer leads to long-chain branching in polyethylene.

Solution

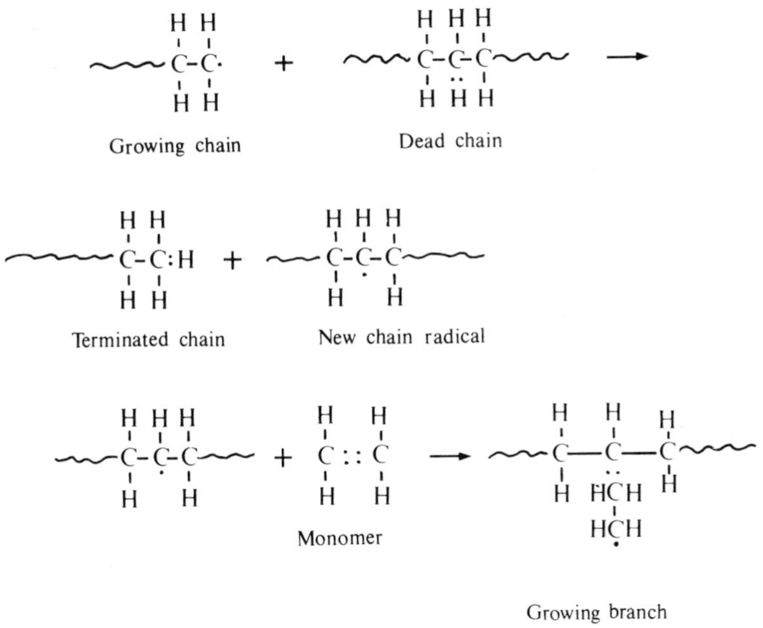

Growing chain   Dead chain

Terminated chain   New chain radical

Monomer   Growing branch

Example 5.   Show, as above, how short branches arise in polyethylene when a growing chain "bites its own back," that is, transfers to an atom a few (5–8) carbon atoms down the chain.

Solution

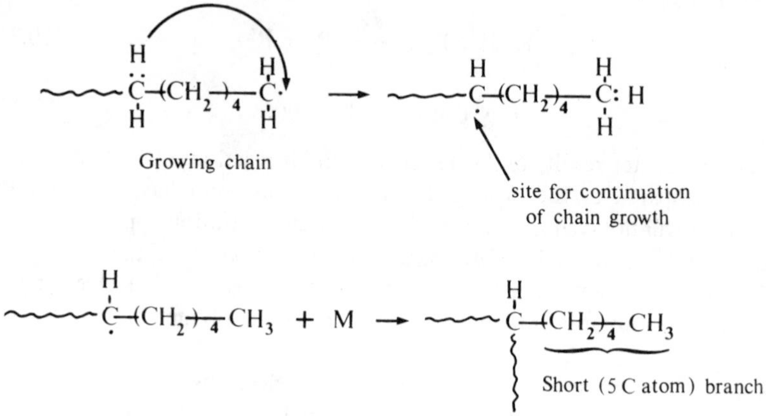

Growing chain

site for continuation
of chain growth

Short (5 C atom) branch

Most frequently, mercaptans (R'S:H) are added to the reaction mass as effective chain-transfer agents to lower the average chain length.

Example 6.   A mercaptan R'S:H is added to the reaction in Example 1. Tabulate the *additional* possibilities for groups A and B and D and E.

Solution.   Chains may now be terminated by the chain-transfer agent, and new ones started by it, giving rise to the additional possibilities

a.

A	B
R'S	H (started by chain transfer, terminated by chain transfer)

	H H
	| |
R'S	—C=C (started by chain transfer, terminated by
	| disproportionation)
	X

b.

D	E
R'S	R'S (both ends started by chain transfer)
R	R'S (one end started by initiator, other by chain transfer)

The rate of dead chain formation by chain transfer

$$r_{tr} = -\left(\frac{d[P\cdot]}{dt}\right)_{tr} = k_{tr}[R':H][P\cdot] \qquad (10.33)$$

must be added to the denominator of (10.22) to give the total rate of dead chain formation

$$\bar{x}_n = \frac{k_p[P\cdot][M]}{(2k_{td} + k_{tc})[P\cdot]^2 + k_{tr}[R':H][P\cdot]} \qquad (10.34)$$

Using (10.11) and (10.21) gives

$$\bar{x}_n = \frac{k_p[M]}{\xi(fk_dk_t[I])^{1/2} + k_{tr}[R':H]} \qquad (10.35)$$

Taking the reciprocal of (10.35)

$$\frac{1}{\bar{x}_n} = \frac{1}{(\bar{x}_n)_0} + C\frac{[R':H]}{[M]} \qquad (10.36)$$

where C is the *chain-transfer constant* $= k_{tr}/k_p$ and $(\bar{x}_n)_0$ is simply the average chain length in the absence of transfer (10.22). Thus a plot of $1/\bar{x}_n$ versus $[R':H]/[M]$ is linear with a slope of C and an intercept of $1/(\bar{x}_n)_0$.

Example 7.   Prove that the use of a chain-transfer agent with $C = 1$ will maintain the ratio $[R':H]/[M]$ constant in a batch reaction.

Solution

$$\frac{r_{tr}}{r_p} = \frac{k_{tr}[R':H][P\cdot]}{k_p[M][P\cdot]} = C\frac{[R':H]}{[M]} = \frac{-d[R':H]/dt}{-d[M]/dt}$$

canceling dt's, separating variables, and integrating from $t = 0$ gives

$$\int_{[R':H]_0}^{[R':H]} \frac{d[R':H]}{[R':H]} = C\int_{[M]_0}^{[M]} d[M]/[M]$$

$$\frac{[R':H]}{[R':H]_0} = \left(\frac{[M]}{[M]_0}\right)^C \qquad (10.37)$$

Therefore, only if $C = 1$ will $[R':H]/[M] = [R':H]_0/[M]_0 =$ constant.

C's greater than 1 use up chain-transfer agent too quickly, leaving nothing to modify the polymer formed at high conversions, while low C's leave a lot of unreacted chain-transfer agent around at the end of the reaction when enough is used to be effective at the beginning. If the agent happens to be a mercaptan, and if you have ever smelled a mercaptan, you can appreciate what a problem that can be. These problems are circumvented by (a) adding a reactive (large C) agent over the course of the reaction, and (b) using mixtures of active and sluggish agents in the initial charge.

Example 8.   Generalize (10.36) to incorporate transfer to a variety of different possible transfer agents S_i

$$P \cdot + S_i \xrightarrow{k_{tr, S_i}} P + S_i \cdot$$

Solution.   The rate of dead chain formation from transfer to each agent S_i must be added to the denominator of (10.22):

$$\bar{x}_n = \frac{k_p [P \cdot] [M]}{(2k_{td} + k_{tc})[P \cdot]^2 + \sum_i k_{tr, S_i} [S_i] [P \cdot]}$$

Invoking (10.11) and (10.21) gives

$$\bar{x}_n = \frac{k_p [M]}{\xi (f k_d k_t [I])^{1/2} + \sum_i k_{tr, S_i} [S_i]}$$

or

$$\frac{1}{\bar{x}_n} = \frac{1}{(\bar{x}_n)_0} + \sum_i C_{S_i} \frac{[S_i]}{[M]}$$

where

$$C_{S_i} = k_{tr, S_i}/k_p.$$

10.7   INSTANTANEOUS DISTRIBUTIONS IN FREE-RADICAL ADDITION POLYMERIZATION

First, consider only chains whose growth is terminated by *either* disproportionation ($\xi = 2$) or chain transfer. The resulting distributions are the same because a growing chain doesn't know if it has been killed by another growing chain or a chain-transfer agent.

The probability q that a growing chain will add a monomer unit (rather than transfer or disproportionate) is

$$q = \frac{\text{rate of propagation}}{\text{rate of propagation} + \text{rate of transfer} + \text{rate of disproportionation}}$$

$$= \frac{k_p [M] [P \cdot]}{k_p [M] [P \cdot] + k_{tr}[P \cdot] [R':H] + 2k_{td}[P \cdot]^2} \tag{10.38}$$

which becomes, with the aid of (10.11),

$$q = \frac{k_p [M]}{k_p [M] + k_{tr}[R':H] + 2(f k_d k_{td} [I])^{1/2}} \tag{10.39}$$

By taking the reciprocal of (10.39)

$$\frac{1}{q} = 1 + \frac{k_{tr}[R':H]}{k_p[M]} + \frac{2(fk_d k_{td}[I])^{1/2}}{k_p[M]} \qquad (10.40)$$

$$\underbrace{\qquad\qquad\qquad\qquad\qquad}$$

$$\frac{1}{\bar{x}_n} \quad (\xi = 2) \qquad (10.35)$$

Therefore

$$\bar{x}_n = \frac{q}{1-q} \qquad (10.41)$$

Now, a polymer molecule containing x units, P_x, has been formed by $(x-1)$ propagation steps, each with a probability q, and one disproportionation or transfer step with a probability $(1-q)$. The probability of finding such a molecule is equal to its number (mole) fraction; therefore

$$\frac{n_x}{N} = (1-q)q^{(x-1)} \qquad (10.42)$$

which, lo and behold, is the "most probable" distribution (9.3) again. But although (10.42) and (9.3) appear the same, p and q are totally different quantities. In batch condensation polymerization, p increases monotonically from 0 to 1 as the reaction proceeds. In free-radical addition, however, q depends only indirectly on conversion, since $[I]$, $[M]$, and $[R'H]$ will, in general, vary with conversion. In the usual case, q is always close to 1, as indeed it must be, to form high-molecular-weight polymer.

Example 9.   Calculate q and $\bar{x}_n$ for the styrene polymerization in Example 2 *for conditions at the start of the reaction.* Neglect chain transfer and assume that $k_t = k_{td}$ (i.e., $\xi = 2$).

Solution.   From (10.39), with $k_{tr} = 0$

$$q = \frac{1}{1 + \{2(fk_d k_{td}[I])^{1/2}/k_p[M]\}}$$

At the start of the reaction, $[M] = [M_0] = 8.7\,g\,moles/liter$ and $[I] = [I_0] = 0.05\,g\,moles/liter$. With the rate constants from Example 2

$$q = \frac{1}{1 + (2/8.7)[(1 \times 0.96 \times 10^{-5} \times 0.05)/(1.18 \times 10^{-3})]^{1/2}} = 0.9954$$

$$\bar{x}_n = \frac{q}{(1-q)} = 216$$

Note that according to (9.6), $\bar{x}_n$ for the most probable distribution should be

$$\bar{x}_n = \frac{1}{1-q} \tag{10.43}$$

which is greater by unity than the value given by (10.41). This discrepancy arises from the fact that the kinetic derivation neglects the monomer molecule added to the chain in the addition step, while that molecule is considered in the derivation of (10.42). From a practical standpoint, this makes little difference, because in the usual free-radical addition polymerization $q \to 1$ and $\bar{x}_n \gg 1$.

By analogy to (9.12)–(9.14) for condensation polymerization

$$\frac{W_x}{W} = xq^{(x-1)}(1-q)^2 \tag{10.44}$$

$$\bar{x}_w = \frac{1+q}{1-q} \tag{10.45}$$

and

$$\frac{\bar{x}_w}{\bar{x}_n} = 1+q \tag{10.46}$$

Here, however, unlike the case in polycondensation, the approximations $q \to 1$, $\bar{x}_n \gg 1$, and $\ln q \cong (q-1)$ are almost always very good, and (10.42), (10.44), and (10.46) simplify to

$$\frac{n_x}{N} \cong \frac{1}{\bar{x}_n} e^{-x/\bar{x}_n} \tag{10.47}$$

$$\frac{W_x}{W} \cong \frac{x}{(\bar{x}_n)^2} e^{-x/\bar{x}_n} \tag{10.48}$$

and

$$\frac{\bar{x}_w}{\bar{x}_n} \cong 2 \tag{10.49}$$

For chains that terminate *exclusively* by combination ($\xi = 1$),

$$q = \frac{k_p[M]}{k_p[M] + 2(fk_d k_{tc}[I])^{1/2}} \tag{10.50}$$

$$\frac{1}{q} = 1 + \underbrace{\frac{2(fk_d k_{tc}[I])^{1/2}}{k_p[M]}}_{} \tag{10.51}$$

$$\frac{2}{\bar{x}_n}(\xi = 1) \qquad (10.35)$$

from which

$$\bar{x}_n = \frac{2q}{(1-q)} \tag{10.52}$$

A dead chain consisting of x units is formed by the combination of two growing chains, one containing y units, and the other $(x - y)$ units:

$$P_y \cdot + P_{(x-y)} = P_x \qquad (10.53)$$

One growing chain involves $(y - 1)$ propagation steps, each of probability q. The other growing chain involves $(x - y - 1)$ propagation steps, each of probability q. Two growing chains are terminated, each with a probability $(1 - q)$. But each dead chain consisting of x total units could have been formed in $(x - 1)$ different ways, for example,

x	Possible Combinations of $(y) + (x - y)$
2	$1 + 1$
3	$2 + 1, 1 + 2$
4	$3 + 1, 2 + 2, 1 + 3$
5	$4 + 1, 3 + 2, 2 + 3, 1 + 4$
$\vdots$	$\vdots$

and all possible ways to form a chain of x units total must be counted. Therefore,

$$\frac{n_x}{N} = (x - 1)q^{(y-1)}q^{(x-y-1)}(1 - q)^2 = (x - 1)(1 - q)^2 q^{(x-2)} \qquad (10.54)$$

For this distribution,

$$\bar{x}_n = \frac{2}{1 - q} \qquad (10.55)$$

which is greater by 2 than the value given by (10.52), for the reasons noted above, and

$$\bar{x}_w = \frac{2 + q}{1 - q} \qquad (10.56)$$

$$\frac{\bar{x}_w}{\bar{x}_n} = \frac{2 + q}{2} \qquad (10.57)$$

For the usual free-radical case, where $q \to 1$, $\bar{x}_n \gg 1$, and $\ln q \cong (q - 1)$,

$$\frac{n_x}{N} \cong \frac{4x}{\bar{x}_n^2} e^{-2x/\bar{x}_n} \qquad (10.58)$$

$$\frac{w_x}{W} \cong \frac{4x^2}{\bar{x}_n^3} e^{-2x/\bar{x}_n} \qquad (10.59)$$

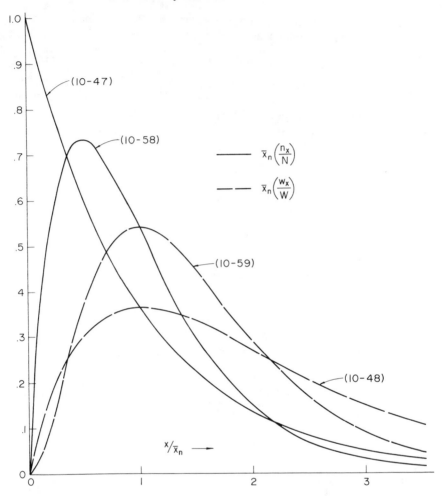

Figure 10.3 Instantaneous number- and weight-fraction distributions of chain lengths in free-radical addition polymerization. Equations (10.47) and (10.48), termination by disproportionation and/or chain transfer; (10.58) and (10.59), termination by combination.

and

$$\frac{\bar{x}_w}{\bar{x}_n} \cong 1.5 \tag{10.60}$$

These distributions are plotted in reduced form in Fig. 10.3. Note that termination by combination sharpens the distributions because of the low probability of two very short or two very long macroradicals combining.

Example 10. Show the structures represented by $x = 1$ in (10.42) for polystyrene. Assume initiation according to

$$R:R \rightarrow 2R\cdot$$

and no chain transfer.

Solution. Since chains are killed only by disproportionation, *dead* chains of length $x = 1$ would be

This points out another important difference between the seemingly similar equations (10.42) and (9.3). The distributions for free-radical addition include only *terminated chains*, *not* unreacted monomer. In polycondensation, there are *no* terminated chains short of complete conversion, and $x = 1$ in (9.3) represents unreacted monomer.

As shown above, chains terminated by combination follow a different distribution than those terminated by disproportionation or chain transfer. If a material contains chains formed according to both distributions (as would happen, for example, if a chain-transfer agent were added to a system that terminates inherently by combination), the distributions must be summed according to the proportion (mole or mass fraction) of each, as detailed in reference 2. Let ψ be the mole (number) fraction of the chains that have been terminated by combination. The weight fraction of chains terminated by combination is $2\psi/(\psi + 1)$. In terms of kinetic parameters, ψ is given by

$$\psi = \left[\frac{2 - \xi}{\xi + \{k_{tr}[R':H]/(k_t f k_d [I])^{1/2}\}} \right] \quad (10.61)$$

where ξ is defined in (10.21). In terms of ψ, the general distributions become

$$\bar{x}_n\left(\frac{n_x}{N}\right) = \left[\psi(\psi + 1)^2 \left(\frac{x}{\bar{x}_n}\right) + (1 - \psi^2) \right] \exp\left[-(\psi + 1)\left(\frac{x}{\bar{x}_n}\right) \right] \quad (10.62)$$

$$\bar{x}_n\left(\frac{w_x}{W}\right) = \left[\psi(\psi + 1)^2 \left(\frac{x}{\bar{x}_n}\right) + (1 - \psi^2) \right] \left(\frac{x}{\bar{x}_n}\right) \exp\left[-(\psi + 1)\left(\frac{x}{\bar{x}_n}\right) \right] \quad (10.63)$$

$$\frac{\bar{x}_w}{\bar{x}_n} = \frac{4\psi + 2}{(\psi + 1)^2} \quad (10.64)$$

For termination by disproportionation and/or chain transfer, $\psi = 0$, and (10.62)–(10.64) reduce to (10.47)–(10.49); for termination exclusively by combination, $\psi = 1$, and they reduce to (10.58)–(10.60), but they are also able to describe various degrees of mixed termination.

10.8   INSTANTANEOUS QUANTITIES

It is obvious from (10.12) that the rate of polymerization is an instantaneous quantity; it depends on the particular values of [M], [I], and T (through the temperature dependence of the rate constants) that exist *at a particular instant* (and location in a reactor, for that matter). In a uniform isothermal batch reaction (Example 2), it decreases monotonically because of the decreases in both [M] and [I] with time. In a similar fashion, $\bar{x}_n$ (10.35) is a function of [M], [I], T, *and* [R′:H], all of which may vary with time (and/or location) in a reactor. But is the concept of an *instantaneous* $\bar{x}_n$ valid? How much do these quantities change during the lifetimes of individual chains? This important point is clarified in Example 11.

Example 11.   Consider the uniform isothermal batch polymerization of Example 2. For conditions *at the beginning of the reaction* ($X = 0$), calculate

a.   the average lifetime of a growing chain, $\bar{t}$,

b.   the percent decrease in initiator concentration in the interval $\bar{t}$,

c.   the conversion in the time interval, $\bar{t}$, and

d.   the number of growing chains per liter of reaction mass.

Additional necessary data: $k_p = 176$ liter/mole sec.

Solution

a.   This is always conceptually difficult. Think as if you were calculating the time required for a car traveling at a constant velocity to cover a specified distance. Here the distance is the kinetic chain length ν (10.17), the average number of monomer units added to a chain during its lifetime (note that the mode of termination is immaterial). The velocity is the rate at which monomer is added to a *single* chain. Monomer is added to *all* chains (in a unit volume of reaction mass) at a rate r_p. In that unit volume of reaction mass, there are [P·] growing chains. Therefore the rate of monomer addition to a single chain $= r_p/[P\cdot]$:

$$\bar{t} = \frac{\nu[P\cdot]}{r_p}$$

With (10.11), (10.12), and (10.17), this becomes

$$\bar{t} = \frac{1}{2(fk_d k_t [I_0])^{1/2}} = \frac{1}{2[1 \times 0.96 \times 10^{-5} \times (176^2/1.18 \times 10^{-3}) \times 0.05]^{1/2}}$$

$$= 0.141 \quad \sec(!)$$

b.   From (10.14b)

$$\frac{[I]}{[I_0]} = \exp(-k_d t) = \exp(-0.96 \times 10^{-5} \times 0.141) = 0.99999865$$

$$\% \text{ decrease } = \left(1 - \frac{[I]}{[I_0]}\right) \times 100 = 1.35 \times 10^{-4}\%$$

c.   The result of b justifies the use of (10.13):

$$\ln \frac{[M]}{[M_0]} = -k_p \left(\frac{fk_d [I_0]}{k_t}\right)^{1/2} t$$

$$= -(1.18 \times 10^{-3} \times 0.96 \times 10^{-5} \times 0.05)^{1/2} \times 0.141$$

$$= -3.36 \times 10^{-6}$$

$$\frac{[M]}{[M_0]} = 0.99999664$$

$$X = \left(1 - \frac{[M]}{[M_0]}\right) = 3.36 \times 10^{-6} = 3.36 \times 10^{-4}\%$$

d.

$$[P\cdot] = \left(\frac{fk_d [I]}{k_t}\right)^{1/2} = \left(\frac{1 \times 0.96 \times 10^{-5} \times 0.05}{(176)^2/1.18 \times 10^{-3}}\right)^{1/2}$$

$$= 1.35 \times 10^{-7} \frac{\text{g moles growing chains}}{\text{liter}}$$

$$1.35 \times 10^{-7} \frac{\text{g moles growing chains}}{\text{liter}} \times 6.02 \times 10^{23} \frac{\text{growing chains}}{\text{g mole}}$$

$$= 8.14 \times 10^{16} \frac{\text{growing chains}}{\text{liter}}$$

Some very important conclusions can be drawn from the preceding example. In a typical homogeneous, free-radical addition polymerization, there are lots of chains growing at any instant. The average lifetime of a growing chain, however, is extremely short — many orders of magnitude smaller than the half-lives of either monomer or initiator. Once a chain has been initiated by the decomposition of an initiator molecule (recall that this is the slow step), it grows and dies in a flash, and once terminated, it plays no further role in the reaction (unless it happens to act as a chain-transfer agent, Example 4) and merely sits around inertly as more chains are formed. This is to be contrasted with condensation polymerization, in which the chains always maintain their terminal reactivity and continue to grow throughout the reaction. The lifetime of free-radical chains is, in fact, so short that changes in concentrations are entirely negligible during a chain lifetime. Hence it is perfectly proper to characterize chains formed at any instant when $[M]$, $[I]$, T, and $[R':H]$ have a particular set of values. All the quantities defined to this point: $\bar{x}_n$, $\bar{x}_w$, q, and therefore the distributions, are just such *instantaneous* properties. For this reason, it was necessary to specificy conditions "at the start of the reaction" in Examples 9 and 11 to permit their calculation.

10.9   CUMULATIVE QUANTITIES

Unfortunately (for the sake of simplicity), as conditions vary within a polymerization reactor, so do the instantaneous quantities. The polymer in the reactor is a mixture of material formed under varying conditions of temperature and concentrations, and therefore must be characterized by *cumulative* quantities, which are integrated averages of the instantaneous quantities of the material formed up until the reactor is sampled. The cumulative number-average chain length $\langle \bar{x}_n \rangle$ is simply the total number of moles of monomer polymerized over the total number of moles of dead chains formed. In a batch reactor, for example,

$$\text{moles of monomer in dead chains } = [M_0] - [M](t)$$

$$\text{moles of dead chains formed } = f\xi([I_0] - [I](t)) + ([R':H_0] - [R':H](t))$$

(For termination by disproportionation, each initiator molecule results in two dead chains, and for combination, one; hence the ξ. Each reacted molecule of chain-transfer agent results in one dead chain.) Therefore

$$\langle \bar{x}_n \rangle(t) = \frac{[M_0] - [M](t)}{f\xi([I_0] - [I](t)) + ([R':H_0] - [R':H](t))} \qquad (10.65)$$

Note that the preceding expression is indeterminate at $t = 0$. Under these conditions, however, $\bar{x}_n = \langle \bar{x}_n \rangle$; both are given by (10.35).

Example 12.   For the reaction in Example 2, with $\xi = 2$, calculate and plot $\bar{x}_n$ and $\langle \bar{x}_n \rangle$ versus X (conversion). Neglect chain transfer.

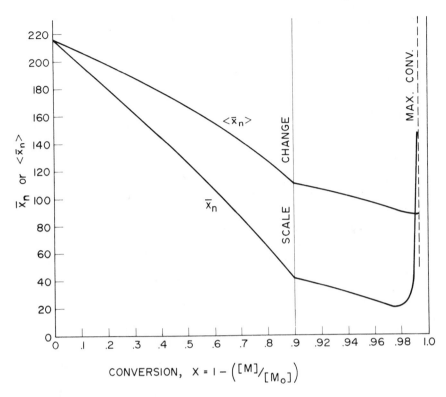

Figure 10.4   Instantaneous and cumulative number-average chain lengths versus conversion for an isothermal, free-radical, batch polymerization (data of Example 12).

Solution.   Calculations here are facilitated by taking time as the independent variable. For this uniform isothermal batch reaction, [M] and [I] are given as functions of time by (10.15) and (10.14b). X versus t is obtained from (10.15) as in Example 2. $\bar{x}_n$ is calculated from (10.35) with $k_{tr} = 0$. If it were necessary to include chain transfer, [R':H] would be calculated from (10.37).

The results are shown in Fig. 10.4. Note that in this case $\bar{x}_n$ goes through a minimum at high conversions. In fact, it ultimately goes to infinity, because in this example of "dead-stop" polymerization [I] goes to zero while [M] remains finite. Not so obvious on this scale is the fact that $\langle\bar{x}_n\rangle$ must also go through a minimum. As long as $\bar{x}_n$ is less than $\langle\bar{x}_n\rangle$, the former will continue to drag the latter down. When $\bar{x}_n$ is greater, it must pull $\langle\bar{x}_n\rangle$ up. Therefore, there *must* be a minimum in $\langle\bar{x}_n\rangle$ where the two curves cross, although $\langle\bar{x}_n\rangle$ does reach a finite asymptote of 86.4 at the maximum conversion of 99.3%.

It is instructive to compare these results with (9.6) for polycondensation. There, $\bar{x}_n$ increases monotonically with conversion (p), and very high conversions *are necessary* for high chain lengths. This is not true for free-radical addition, where chains formed early in the reaction have fully developed chain lengths.

Keep in mind also that if samples were removed from the reactor and analyzed for number-average chain length, the quantity determined would be $\langle \bar{x}_n \rangle$. The *cumulative* quantity is what characterizes the reactor contents. The only way to measure $\bar{x}_n$ would be to sample at a very low conversion, where $\bar{x}_n \cong \langle \bar{x}_n \rangle$.

10.10   RELATIONS BETWEEN INSTANTANEOUS AND CUMULATIVE AVERAGE CHAIN LENGTHS

Given either the instantaneous or cumulative average chain length as a function of conversion, the other may be calculated as follows:

Recall that conversion may be written in terms of weights or concentrations (assuming constant volume for the latter) of unreacted monomer:

$$X = \frac{W_0 - W}{W_0} = \frac{[M_0] - [M]}{[M_0]}$$

moles of monomer polymerized per unit volume up to conversion X:

$$[M_0] - [M] = [M_0] X$$

moles of monomer polymerized per unit volume in conversion increment dX:

$$d[M] = [M_0] dX$$

moles of polymer chains formed per unit volume in conversion increment dX:

$$dN = \frac{d[M]}{\bar{x}_n} = \frac{[M_0] dX}{\bar{x}_n}$$

The total moles of polymer chains per unit volume formed *up to* conversion X is obtained by integrating dN:

$$N = \int_0^X dN = \int_0^X \frac{[M_0] dX}{\bar{x}_n}$$

But, by definition of $\langle \bar{x}_n \rangle$, N is also given by

$$N = \frac{\text{moles of monomer polymerized per unit volume to conversion } X}{\langle \bar{x}_n \rangle}$$

$$= \frac{[M_0] X}{\langle \bar{x}_n \rangle}$$

Equating the two expressions for N gives

$$\langle \bar{x}_n \rangle = \frac{X}{\int_0^X dX/\bar{x}_n} \tag{10.66}$$

Differentiating (10.66) reverses the result:

$$\bar{x}_n = \frac{dX}{d(X/\langle \bar{x}_n \rangle)} \tag{10.67}$$

Equation (6.6) gives the weight average of a mixture in terms of a summation of the weight averages of the discrete components of the mixture. For a mixture of differential components of weight-average chain length $\bar{x}_w$, this may be generalized to

$$\langle \bar{x}_w \rangle = \frac{\int_0^X \bar{x}_w \, dW_p}{\int_0^X dW_p} = \frac{1}{W_p} \int_0^X \bar{x}_w \, dW_p$$

where
$$dW_p = \text{weight of polymer formed in a conversion increment } dX$$

$$W_p = \text{total weight of polymer formed up to conversion } X$$

Now $W_p = W_0 X$ and $dW_p = W_0 dX$; so

$$\langle \bar{x}_w \rangle = \frac{1}{X} \int_0^X \bar{x}_w \, dX \tag{10.68}$$

By differentiating (10.68),

$$\bar{x}_w = \langle \bar{x}_w \rangle + X \left(\frac{d\langle \bar{x}_w \rangle}{dX} \right) \tag{10.69}$$

The relation between $\bar{x}_n$ and $\bar{x}_w$ is, of course, determined by the nature of the instantaneous distribution, which fixes the *instantaneous* polydispersity index, $\bar{x}_w/\bar{x}_n$, through (10.64). However, it is not the instantaneous polydispersity index that characterizes the reactor product, but rather the *cumulative* polydispersity index $\langle \bar{x}_w \rangle / \langle \bar{x}_n \rangle$. Even though the instantaneous polydispersity index may remain constant, if the instantaneous averages vary during the reaction (as in Example 12), the cumulative distribution *must* be broader than the instantaneous, increasing the cumulative polydispersity index (Chapter 6, Example 2, provides a simple quantitative illustration of this). This is one of the reasons why commercial polymers often have polydispersity indices much greater than the minimum value given by (10.64).

Example 13.   Which type of isothermal reactor would produce the narrowest possible distribution of chain lengths in a free-radical addition polymerization:

continuous stirred tank (backmix), batch (assume perfect stirring in each of the previous), plug-flow tubular, or laminar-flow tubular?

Solution.   Only in an ideal continuous stirred tank reactor are [M] and [I], and therefore $\bar{x}_n$, constant, giving the narrowest possible distribution, due only to the microscopically random nature of the reaction. In a batch reactor [M] and [I] vary with time, and in tubular reactors, with position, thus causing a shift in $\bar{x}_n$ with conversion, broadening the distribution.

Example 14.   Consider the isothermal, free-radical, batch polymerization of a monomer with $[M_0] = 1.6901$ g moles/liter and $[I_0] = 1.6901 \times 10^{-3}$ g moles/liter. At the reaction temperature, $(k_p^2/k_t) = 26.1$ liter/mole sec and $k_d = 4.369 \times 10^{-7}$ sec^{-1}. This particular polymer terminates by disproportionation ($\xi = 2$). Neglect chain transfer. Assume perfect reactor stirring and $f = 1$, and calculate and plot $\bar{x}_n$, $\langle \bar{x}_n \rangle$, $\bar{x}_w$, $\langle \bar{x}_w \rangle$, and $\langle \bar{x}_w \rangle / \langle \bar{x}_n \rangle$ as functions of conversion for this system. Hint: for this particular system (k_p^2/k_t) is very large and k_d is small (compare with Example 2). This has some important consequences. First, the chain lengths will be tremendous. (In real life, it is likely that chain lengths here would be transfer limited; that is, the growing chains would transfer to *something* in the reaction mass before reaching the lengths calculated.) Second, [I] remains essentially constant throughout the course of the reaction. As a result, $\bar{x}_n$ is linear in X (combine (10.22) with the definition of conversion to see why), which in turn allows easy analytical evaluation of (10.66)–(10.69)(but also renders (10.65) useless).

Solution.   Following the suggestion in the hint,

$$\bar{x}_n = \frac{k_p[M]}{\xi(fk_dk_t)^{1/2}[I]^{1/2}} = \frac{k_p[M_0]}{\xi(fk_dk_t)^{1/2}[I]^{1/2}}(1-X) = 158,900(1-X)$$

Plugging this relation into (10.66),

$$\langle \bar{x}_n \rangle = \frac{158,900X}{\int_0^X dX/(1-X)} = \frac{158,900X}{-\ln(1-X)}$$

Now, for $\xi = 2$, $\psi = 0$, $\bar{x}_w = 2\bar{x}_n$ (10.49) or (10.64),

$$\bar{x}_w = 317,800(1-X)$$

Plugging the above into (10.68)

$$\langle \bar{x}_w \rangle = \frac{1}{X}\int_0^X \bar{x}_w \, dX = \frac{317,800}{X}\int_0^X (1-X)dX = 317,800\left(1 - \frac{X}{2}\right)$$

Plots based on these equations are shown in Fig. 10.5. Note that rather high

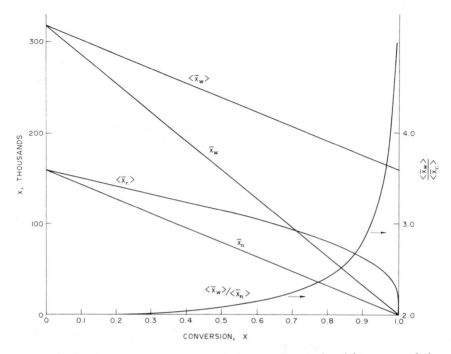

Figure 10.5 Instantaneous and cumulative number- and weight-average chain lengths and the cumulative polydispersity index for an isothermal, free-radical, batch polymerization (data of Example 14).

conversions can be achieved before the polydispersity index starts to shoot up significantly above the minimum (instantaneous) value of 2.0.

If it is necessary to maintain minimum polydispersity (albeit with shorter chain lengths), the drift in instantaneous chain length can often be compensated for by varying the rate of addition of chain-transfer agent to the reactor. In the example above, a high initial rate of a very active agent could be reduced as conversion increased, thereby counteracting the decrease in $\bar{x}_n$ (and increase in polydispersity) which would otherwise occur. Adjusting the rate of addition of chain-transfer agent to maintain $\bar{x}_n$ constant has been discussed for batch (3) and continuous (2) reactors.

Example 14 constitutes one of the simplest possible illustrations of what is sometimes termed "polymer reactor engineering." Even a minor complication, such as substitution of the parameters of Example 2 or the inclusion of chain transfer, would necessitate a numerical solution. While the basic principles are there, additional detail is beyond the scope of this chapter, but you might wish to consider the application of these principles to a nonisothermal reactor for example.

10.11   EMULSION POLYMERIZATION

The preceding discussion of free-radical addition polymerization has considered only *homogeneous* reactions. Considerable polymer is produced commercially by a complex *heterogeneous* free-radical addition process known as emulsion polymerization. This process was commercialized in the United States during World War II to provide a source of synthetic rubber. A rational explanation of the mechanism of emulsion polymerization was proposed by Harkins[4] and quantified by Smith and Ewart[5] after the war, when information gathered at various locations could be freely exchanged. Perhaps the best way to introduce the subject is to list a typical reactor charge:

Typical Emulsion Polymerization Charge
100   parts (by weight) monomer (water insoluble)
180   parts water
 2–5   parts fatty acid soap
0.1–0.5   part *water-soluble* initiator
 0–1   part chain-transfer agent (monomer soluble)

The first question is "what's the soap for?" Soaps are the sodium or potassium salts of organic acids or sulfates:

$$\left[R{-}\overset{\overset{\displaystyle O}{\parallel}}{C}{-}O \right]^{-} Na^{+}$$

When they are added to water in low concentrations, they ionize and float around freely much as sodium chloride ions would. The anions, however, consist of a highly polar hydrophilic (water-seeking) "head" and an organic, hydrophobic (water-fearing) "tail." As the soap concentration is increased, a value is suddenly reached where the anions begin to agglomerate in *micelles* rather than float around individually. These micelles have dimensions on the order of 50 to 60 Å (1 cm = 10^4 μm = 10^8 Å), far too small to be seen with a light microscope. They consist of a tangle of the hydrophobic tails in the interior (getting as far away from the water as possible) with the hydrophilic heads on the outside. This process is easily observed by following the variation of a number of solution properties with soap concentration, for example, electrical conductivity or surface tension (Fig. 10.6). The break occurs when micelles start to form and is known as the *critical micelle concentration* or CMC.

When an organic monomer is added to an aqueous micelle solution, it naturally prefers the organic environment at the interior of the micelles. Some of it congregates there, swelling the micelles until an equilibrium is reached with the contraction force of surface tension. *Most* of the monomer, however, is distributed

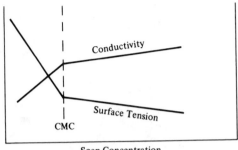

Figure 10.6   Variation of solution properties at the critical micelle concentration.

in the form of *much* larger (1 μm or 10^4 Å) droplets stabilized by soap. This complex mixture is an *emulsion*. The cleaning action of soaps depends on their ability to emulsify oils and greases.

Despite the fact that most of the monomer is present in the droplets, the swollen micelles, because of their much smaller size, present a much larger surface area than the droplets. This is easily seen by assuming a micelle volume to drop volume ratio of 1/10 and using the ballpark figures given above. Since the surface/volume ratio of a sphere is $3/R$,

$$\frac{S_{\text{micelle}}}{S_{\text{droplet}}} = \left(\frac{V_{\text{micelle}}}{V_{\text{droplet}}}\right)\left(\frac{R_{\text{droplet}}}{R_{\text{micelle}}}\right) \cong \left(\frac{1}{10}\right)\left(\frac{10^4}{0.5 \times 10^2}\right) \cong 20$$

Free radicals for emulsion polymerization are generated *in the aqueous phase* by the decomposition of water-soluble initiators, usually potassium or ammonium persulfate:

$$S_2O_8^= \longrightarrow 2SO_4^-$$

Persulfate   Sulfate ion-radical

"Redox" systems, so called because they involve the alternate oxidation and reduction of a trace catalyst, are a newer and more efficient means of generating radicals, for example,

1. $S_2O_8^= + HSO_3^- \rightarrow SO_4^= + SO_4^- + HSO_3$

     Persulfate   Bisulfite

2. $S_2O_8^= + Fe^{++} \rightarrow SO_4^= + Fe^{3+} + SO_4^-$

3. $HSO_3^- + Fe^{3+} \rightarrow HSO_3 + Fe^{++}$

                  ——————————————————

     $S_2O_8^= + H\dot{S}O_3^- \rightarrow SO_4^= + SO_4^- + HSO_3$

The original wartime GR-S (polybutadiene-co-styrene) polymerization was carried out at 50°C using potassium persulfate initiator ("hot" rubber). The use

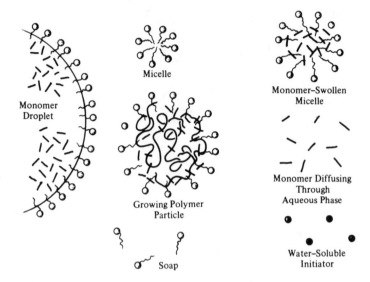

Figure 10.7   Structures in emulsion polymerization.

of the more efficient redox systems allowed a reduction in polymerization temperature to 5°C ("cold" rubber). The latter product has superior properties because the lower polymerization temperature promotes cis 1-4 addition of the butadiene. Figure 10.7 illustrates the structures present during emulsion polymerization.

The radicals thus generated in the aqueous phase bounce around until they encounter some monomer. Since the surface area presented by the monomer-swollen micelles is so much greater than that of the droplets, the probability of a radical entering a monomer-swollen micelle rather than a droplet is large. As soon as the radical encounters the monomer within the micelle, it initiates polymerization. The conversion of monomer to polymer within the growing micelle lowers the monomer concentration therein, and monomer begins to diffuse from uninitiated micelles and monomer droplets to the growing polymer-containing micelles. Those monomer-swollen micelles not struck by a radical during the initial stages of conversion thus disappear, losing their monomer to those that have been initiated, stabilizing the number of growing particles in the system. The reaction mass now consists of a stable number of growing polymer particles (originally micelles) and the monomer droplets. This first phase of the reaction was termed *Stage I* by Smith and Ewart. The reaction now enters *Stage II*, in which the monomer droplets simply act as reservoirs supplying monomer to the growing polymer particles by diffusion through the water. The monomer concentration in the growing particles maintains a nearly constant dynamic equilibrium value dictated by the tendency toward further dilution (increasing the entropy) and the opposing effect of surface tension attempting to minimize the surface area. (Although most organic monomers are normally thought of as being water "insoluble," their concentrations in the aqueous phase, though small, are sufficient to permit a high enough diffusion flux to maintain the monomer concentration in the polymerizing particles.[6]) Smith and Ewart then subdivided Stage II into three subcases. The description that follows applies to their case 2, which most often applies to what happens in the real world.

A monomer-swollen micelle that has been struck by a radical now contains *one* growing chain. With only one radical per particle, there is no way in which the chain can terminate, and it continues to grow until a second initiator radical enters the particle. Under conditions prevailing within the particle in case 2, the rate of termination is much greater than the rate of propagation; so the chain growth is terminated essentially immediately after the entrance of the second initiator radical.[6] The particle then remains dormant until a third initiator radical enters, initiating the growth of a second chain. This second chain grows until it is terminated by the entry of the fourth radical, and so on.

10.12   KINETICS OF EMULSION POLYMERIZATION IN STAGE II, CASE 2

Thus, in Stage II, the reaction mass consists of a stable number of monomer-swollen polymer particles that are the loci of all polymerization. At any given time (for case 2), a particle contains either one growing chain or no growing chains. Statistically, then, if there are N particles per liter of reaction mass, there are $N/2$ growing chains per liter of reaction mass.

The polymerization rate is given by

$$r_p = k_p[\text{M}][\text{P}\cdot]$$

where k_p is the usual *homogeneous* propagation rate constant for polymerization *within* the particles and $[\text{M}]$ is the equilibrium monomer concentration *within a particle*. Now,

$$[\text{P}\cdot] = \frac{N}{2A} \quad \text{moles radicals per liter of } \textit{reaction mass} \text{ (particles plus water)}$$

(10.70)

where A is Avogadro's number (radicals per mole radicals).

The rate of polymerization is then

$$r_p = k_p\left(\frac{N}{2A}\right)[\text{M}]$$

(10.71)

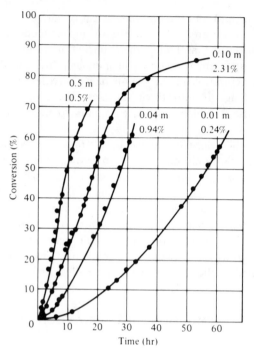

Figure 10.8   The emulsion polymerization of isoprene as a function of soap (potassium laurate) concentration. Reprinted from (4). Copyright 1947 by the AMERICAN CHEMICAL SOCIETY. Reprinted by permission of the copyright owner.

$$\underbrace{\left(\frac{\text{moles monomer}}{\text{liter of reaction mass, sec}}\right)}_{r_p} = \underbrace{\left(\frac{\dfrac{\text{moles monomer}}{\text{liter of particles, sec}}}{\dfrac{\text{moles monomer}}{\text{liter particles}} \times \dfrac{\text{moles radicals}}{\text{liter particles}}}\right)}_{k_p}$$

$$\underbrace{\left(\frac{\text{moles radicals}}{\text{liter reaction mass}}\right)}_{\dfrac{N}{2A}} \underbrace{\left(\frac{\text{moles monomer}}{\text{liter particles}}\right)}_{[M]}$$

Equation (10.71) thus gives the rate of polymerization *per total volume of reaction mass.* Surprisingly, it predicts a rate that is independent of initiator concentration. Moreover, since N is constant in Stage II, as is $[M]$, a constant rate is predicted. This is borne out experimentally as shown in Fig. 10.8.[4]

Deviations from linearity are observed at low conversions as N is being stabilized in Stage I, and at high conversions in Stage III, when the monomer droplets are used up and are no longer able to supply the monomer necessary to maintain [M] constant in the growing particles. Thus in Stage III the rate drops off exponentially as the monomer is exhausted within the particles.

Note the increase in rate with soap concentration. The more soap used, the greater the number of micelles established initially and the higher N will be. The predicted rate independence of initiator concentration must be viewed with caution. It's alright as long as N is held constant, but, in practice, N increases with $[I_0]$. The more initiator added at the start of a reaction, the greater the number of monomer-swollen micelles that start growing before N is stabilized. Methods are available for estimating N.[6-8]

Example 15. The Putrid Paint Division of Crud Chemicals, Inc., has available a 10% (by weight) polyvinyl acetate latex containing 1×10^{14} particles/cm^3. To obtain optimum characteristics as an interior wall paint, a larger particle size and higher concentration of polymer are needed. It is proposed to obtain these by adding an additional 4 parts (by weight) of monomer per part of polymer to the latex and polymerizing without further addition of soap. The reaction will be carried to 85% conversion, and the unreacted monomer will be steam-stripped and recovered. (The conditions of this "seeded" polymerization are set up so that the entire reaction proceeds in Stage II.)

Estimate the time required for the reaction, and the rate of heat removal (Btu/gal of original latex, per hour) necessary to maintain an essentially iso-thermal reaction at 60°C. At what conversion would the monomer droplets disappear and the rate cease to be constant?

Data:
$$k_p = 3700 \frac{\text{liter}}{\text{mole sec}} \text{ at } 60°C$$

$$\Delta H_p = -21 \text{ kcal/mole monomer unit}$$

$$\rho(\text{polymer}) = 1.2 \text{ g/cm}^3, \rho(\text{monomer}) = 0.8 \text{ g/cm}^3$$

The concentration of monomer in growing polymer particles is 10% (by weight).

Solution. It must first be assumed that in the absence of additional soap, no new micelles will be established; so the original latex particles act as the ex-clusive loci for further polymerization. Then [M] and N must be obtained for use in (10.71).

Basis: 100 g monomer-swollen particles, 90 g polymer, 10 g monomer.

$$(90\,\text{g})\left(\frac{1\,\text{cm}^3}{1.2\,\text{g}}\right) = 75\,\text{cm}^3 \text{ polymer}$$

(assumes additivity of volume)

$$(10\,\text{g})\left(\frac{1\,\text{cm}^3}{0.8\,\text{g}}\right) = \frac{12.5\,\text{cm}^3 \text{ monomer}}{87.5\,\text{cm}^3}$$

Molecular weight of vinyl acetate $\left(\begin{array}{c}\text{H H}\\ |\ \ |\\ \text{C=C}\\ |\ \ |\\ \text{H O}\end{array}\right) = 86\,\text{g/g mole}$

$$\begin{array}{c}|\\ \text{C=0}\\ |\\ \text{CH}_3\end{array}$$

$$(10\,\text{g})\left(\frac{1\,\text{g mole}}{86\,\text{g}}\right) = 0.116 \quad \text{g moles monomer present}$$

$$[M] = (0.116\,\text{g moles monomer})\left(\frac{1}{87.5\,\text{cm}^3 \text{ of particle}}\right)\left(\frac{1000\,\text{cm}^3}{\text{liter}}\right)$$

$$= 1.33\,\frac{\text{g moles monomer}}{\text{liter of particles}}$$

$$N = 1 \times 10^{17}\,\frac{\text{particles}}{\text{liter of original latex}}$$

$$\frac{N}{2} = \frac{1 \times 10^{17}}{2}\,\frac{\text{free radicals}}{\text{liter of original latex}}$$

Now, 1 g mole of free radicals = 6×10^{23} free radicals (Avogadro's number):

$$\frac{N}{2A} = \left(\frac{1 \times 10^{17}}{2}\,\frac{\text{free radicals}}{\text{liter original latex}}\right)\left(\frac{1\,\text{g mole free radicals}}{6 \times 10^{23} \text{ free radicals}}\right)$$

$$= \frac{1}{12} \times 10^{-6}\,\frac{\text{g moles free radicals}}{\text{liter original latex}}$$

$$r_p = k_p\,\frac{N}{2A}\,[M]$$

$$= \left(3700\,\frac{\text{liter particles}}{\text{g mole sec}}\right)\left(\frac{1}{12} \times 10^{-6}\,\frac{\text{g moles free radicals}}{\text{liter original latex}}\right)$$

$$\left(1.33 \frac{\text{g moles monomer}}{\text{liter particles}}\right) = 4.1 \times 10^{-4} \frac{\text{g moles monomer}}{\text{liter original latex sec}}$$

$$= 1.48 \frac{\text{g moles monomer}}{\text{liter original latex hr}}$$

By again assuming additivity of volumes, the density of the original latex is found to be $1.02 \, \text{g/cm}^3$. In *1 liter* of the original latex = 1020 g, there are 102 g polymer, to which are added $102 \times 4 = 408$ g monomer, or

$$(408 \text{ g}) \left(\frac{1 \text{ g mole}}{86 \text{ g}}\right) = 4.74 \quad \text{g moles monomer}$$

The reaction converts $4.74 \times 0.85 = 4.02$ g moles monomer which takes

$$\left(\frac{4.02 \text{ g moles monomer}}{\text{liter original latex}}\right)\left(\frac{1 \text{ liter original latex, hr}}{1.48 \text{ g moles monomer}}\right) = 2.72 \text{ hr}$$

For an isothermal reaction, the rate of heat generation = rate of heat removed = $(21,000 \text{ cal/g mole monomer})(1.48 \text{ g moles monomer/liter original latex hr}) = 31,000 \text{ cal/liter original latex hr} = 467 \text{ Btu/gal original latex hr}$. When the monomer droplets just disappear, the reaction enters Stage III, and all the monomer and polymer will be in the swollen polymer particles, a total of 102 g (original polymer) + 408 g (added monomer) = 510 g/liter of original latex. At this point, the particles still contain 10% monomer, so there are $(0.1)(510) = 51$ g of unconverted monomer. Therefore

$$\text{conversion} = \left(1 - \frac{M}{M_0}\right) = 1 - \frac{51}{408} = 0.875 \quad \text{or} \quad 87.5\%$$

Beyond this point, the rate will drop off exponentially as the monomer concentration in the particles falls.

Despite the secondary effect of $[I_0]$ on the rate, it has a strong influence on the average chain length. The greater the rate of radical generation, the greater the frequency of alternation between growth and death in a particle, resulting in a lower chain length. If r_c represents the rate of radical capture per liter of reaction mass (half the radicals of which produce dead chains, in the absence of transfer

$$\bar{x}_n = \frac{k_p(N/2A)[M]}{r_c/2} = k_p \frac{N}{Ar_c}[M] \tag{10.72}$$

The rate of generation of radicals is based on the presence of initiator in the aqueous phase alone:

$$r_{gen}\left(\frac{\text{moles radicals}}{\text{liter aqueous phase sec}}\right) = 2k_d(\text{sec}^{-1})[I]\left(\frac{\text{moles initiator}}{\text{liter aqueous phase}}\right)$$

(10.73)

If a steady state in radicals is assumed,

$$r_c\left(\frac{\text{moles radicals}}{\text{liter reaction mass sec}}\right) = r_{gen}\left(\frac{\text{moles radicals}}{\text{liter aqueous phase sec}}\right)$$

$$\phi_a \left(\frac{\text{liter aqueous phase}}{\text{liter reaction mass}}\right) \qquad (10.74)$$

where ϕ_a = volume fraction aqueous phase. Thus

$$r_c = 2k_d[I]\phi_a = 2k_d[I'] \qquad (10.75)$$

where $[I'] = [I]\phi_a$, the moles initiator per total volume of reaction mass and

$$\bar{x}_n = \frac{k_p N[M]}{2Ak_d[I']} \qquad (10.76)$$

The chain length is, therefore, inversely proportional to the *first* power of initiator concentration (compare with (10.22)).

It must be emphasized that the preceding is an oversimplified view of an extremely complex process. Smith and Ewart point out that even within Stage II their case 2 was merely the middle of a spectrum. At one end of the spectrum, case 1, the number of radicals per particle is much less than $\frac{1}{2}$. This is believed to occur sometimes because of radical escape from the particles. At the other extreme, the number of radicals per particle is much greater than $\frac{1}{2}$. This comes about from large particles and/or small k_t. Under these conditions each particle acts as a tiny homogeneous reactor, according to the kinetics and mechanism previously developed for homogeneous free-radical addition. The transition from case 2 to case 3 kinetics is often observed in seeded emulsion polymerizations, where the particles are fairly large to begin with and grow from there as the reaction proceeds. Smith–Ewart theory was developed for such monomers as styrene, with very low water solubility. Monomers such as acrylonitrile, with appreciable water solubility (on the order of 10%), may undergo significant homogeneous initiation in the aqueous phase. In some emulsion systems the particles flocculate (coalesce) during polymerization, not only making a kinetic description difficult, but sometimes badly fouling reactors. Good reviews of the subject are available,[8,9] as well as complete books.[7,10,11]

REFERENCES

1. O. Levenspeil, *Chemical Reaction Engineering*, 2nd ed., Wiley, New York, 1972.

2. T. Kenat, R. I. Kermode and S. L. Rosen, *J. Appl. Polym. Sci.*, **13**, 1353 (1969).

3. B. F. Hoffman, S. Schreiber and G. Rosen, *Ind. Eng. Chem.*, **56**, 5, 51 (1964).

4. W. D. Harkins, *J. Am. Chem. Soc.*, **69**, 1428 (1947).

5. W. V. Smith and R. H. Ewart, *J. Chem. Phys.*, **16**, 6, 592 (1948).

6. P. J. Flory, *Principles of Polymer Chemistry*, Cornell University Press, Ithaca, New York, 1953, Chapter 3.

7. F. A. Bovey et al., *Emulsion Polymerization*, Interscience, New York, 1955.

8. J. L. Gardon, "Emulsion Polymerization," *Polymerization Processes*, C. E. Schildknecht and I. Skeist, Eds., Wiley, New York, 1977, Chapter 6.

9. J. Ugelstad and F. K. Hansen, *Rubber Chem. Technol.*, **49**, 536 (1976).

10. I. Piirma and J. L. Gardon, Eds., *Emulsion Polymerization*, Am. Chem. Soc. Symposium Series 24, American Chemical Society, Washington, 1976.

11. D. C. Blackley, *Emulsion Polymerization*, Wiley, New York, 1975.

CHAPTER **11**

Non-radical Addition Polymerization

11.1 CATIONIC POLYMERIZATION[1,2]

Strong Lewis acids, that is, electron acceptors, are often capable of initiating the addition polymerization of monomers with electron-rich substituents adjacent to the double bond. Cationic catalysts are most commonly metal trihalides such as $AlCl_3$ or BF_3. These compounds, although electrically neutral, are two electrons short of having a complete valence shell of eight electrons. They were found to require traces of a cocatalyst, usually water, to initiate polymerization, first by grabbing a pair of electrons from the cocatalyst:

$$
\begin{array}{cccc}
\text{F} & \text{H} & \text{F} \ \text{H} & \text{F} \\
\text{F} : \ddot{\text{B}} \ + \ : \ddot{\text{O}} : & \longrightarrow & \text{F} : \ddot{\text{B}} : \ \ddot{\text{O}} : & \longrightarrow \left[\text{F} : \ddot{\text{B}} : \ddot{\text{O}} : \text{H} \right]^{-} + \ [\text{H}]^{+} \\
\text{F} & \text{H} & \text{F} \ \text{H} & \text{F}
\end{array}
$$

The leftover proton is thought to be the actual initiating species, abstracting a pair of electrons from the monomer and leaving a cationic chain end which reacts with additional monomer molecules.

$$
[\text{BF}_3\text{OH}]^{-} \ [\text{H}]^{+} + \begin{array}{c} \text{H} \ \ \text{CH}_3 \\ \text{C} :: \text{C} \\ \text{H} \ \ \text{CH}_3 \end{array} \longrightarrow \left[\text{H} : \overset{\text{H}}{\underset{\text{H}}{\text{C}}} : \overset{\text{CH}_3}{\underset{\text{CH}_3}{\text{C}}} \right]^{+} + [\text{BF}_3\text{OH}]^{-} \qquad \text{etc.}
$$

Gregen or counter ion

An important point here is that the *gegen* or *counter* ion is electrostatically held near the growing chain end, and so can exert a steric influence on the addition of

152

monomer units. Termination is thought to occur by a disproportionation-like reaction which regenerates the catalyst complex. The complex, therefore, is a true catalyst, unlike free-radical initiators.

$$\sim\sim\overset{\overset{\textstyle CH_3}{\underset{\textstyle |}{}}}{\underset{\overset{\textstyle |}{\textstyle CH_3}}{\underset{|}{\overset{H}{\underset{|}{\overset{|}{C}}}}-\overset{|}{\underset{|}{C}}}}\Bigg]^+ \;+\; [BF_3OH]^- \;\longrightarrow\; \sim\sim\overset{\overset{\textstyle CH_3}{\underset{\textstyle |}{}}}{\underset{\overset{|}{H}\;\underset{\textstyle CH_2}{\overset{\|}{C}}}{\underset{|}{\overset{H}{\underset{|}{\overset{|}{C}}}}-\overset{|}{\underset{|}{C}}}} \;+\; BF_3 \cdot H_2O$$

The kinetics of these reactions is not well understood, but they proceed very rapidly at extremely low temperatures. For example, the polymerization of isobutylene illustrated above is carried out commercially at $-150°F$. The average chain length increases as the temperature is lowered.

Cationic initiation is successful only with monomers like isobutylene that have electron-rich substituents adjacent to the double bond:

$$\begin{array}{c} H\;\;H \\ |\;\;\;| \\ C=C \\ |\;\;\;| \\ H\;\;O \\ |\\ R \end{array} \qquad\qquad \begin{array}{c} \;\;\;\;CH_3 \\ H\;\;| \\ |\;\;\; \\ C=C \\ |\; \\ H \\ \bigcirc \end{array}$$

Alkyl vinyl ethers α-methyl styrene

None of these monomers can be polymerized to high molecular weight with free-radical initiators.

11.2 ANIONIC POLYMERIZATION[3]

Addition polymerization may also be initiated by anions. Anionic polymerization has achieved tremendous commercial importance in the past two decades because of its ability to control molecular structure during polymerization, allowing the synthesis of materials that were previously difficult or impossible to obtain. A variety of anionic initiators has been investigated, but the organic alkali—metal salts are perhaps most common, as illustrated below for the polymerization of styrene with n-butyllithium

$$\left[\begin{array}{c} H\;H\;H\;H \\ |\;\;|\;\;|\;\;| \\ H-C-C-C-C: \\ |\;\;|\;\;|\;\;| \\ H\;H\;H\;H \end{array}\right]^{-}[Li]^{+} \;+\; \begin{array}{c} H\;H \\ |\;\;| \\ C=C \\ |\;\;| \\ H\;\bigcirc \end{array} \;\longrightarrow\; \left[\begin{array}{c} H\;H\;H\;H\;H\;H \\ |\;\;|\;\;|\;\;|\;\;|\;\;| \\ H-C-C-C-C:C:C: \\ |\;\;|\;\;|\;\;|\;\;|\;\;| \\ H\;H\;H\;H\;H\bigcirc \end{array}\right]^{-}[Li]^{+}$$

N-butyllithium (BuLi)   Styrene

The anionic chain end then propagates the chain by adding another monomer molecule. Again, the gegen ion can sterically influence the reaction.

Sodium and lithium *metals* were used to polymerize butadiene in Germany during World War II. After the war, in the United States, it was discovered that under appropriate conditions lithium could lead to largely *cis*-1,4 addition of butadiene and isoprene (the latter being the synthetic counterpart of natural rubber). In these processes, a metal atom first reacts with the monomer to form an anion radical:

$$
\text{Li} \cdot + \; \overset{\text{H}}{\underset{\text{H}}{\text{C}}} :: \overset{\text{H}}{\text{C}} :: \overset{\text{H}}{\text{C}} :: \overset{\text{H}}{\underset{\text{H}}{\text{C}}} \longrightarrow \left[\cdot \overset{\text{H}}{\underset{\text{H}}{\text{C}}} : \overset{\text{H}}{\text{C}} :: \overset{\text{H}}{\text{C}} : \overset{\text{H}}{\underset{\text{H}}{\text{C}}} : \right]^{-} [\text{Li}]^{+}
$$

Anion radical

These anion radicals then react in either of two ways. One may react with another atom of lithium,

$$
\text{Li} \cdot + \left[\cdot \overset{\text{H}}{\underset{\text{H}}{\text{C}}} : \overset{\text{H}}{\text{C}} :: \overset{\text{H}}{\text{C}} : \overset{\text{H}}{\underset{\text{H}}{\text{C}}} : \right]^{-} [\text{Li}]^{+} \longrightarrow {}^{+}[\text{Li}] \left[: \overset{\text{H}}{\underset{\text{H}}{\text{C}}} : \overset{\text{H}}{\text{C}} :: \overset{\text{H}}{\text{C}} : \overset{\text{H}}{\underset{\text{H}}{\text{C}}} : \right]^{-} [\text{Li}]^{+}
$$

and/or two may rapidly undergo radical recombination

$$
2 \left[\cdot \overset{\text{H}}{\underset{\text{H}}{\text{C}}} : \overset{\text{H}}{\text{C}} :: \overset{\text{H}}{\text{C}} : \overset{\text{H}}{\underset{\text{H}}{\text{C}}} : \right]^{-} [\text{Li}]^{+} \longrightarrow {}^{+}[\text{Li}] \left[: \overset{\text{H}}{\underset{\text{H}}{\text{C}}} : \overset{\text{H}}{\text{C}} = \overset{\text{H}}{\text{C}} : \overset{\text{H}}{\underset{\text{H}}{\text{C}}} : \overset{\text{H}}{\underset{\text{H}}{\text{C}}} : \overset{\text{H}}{\text{C}} = \overset{\text{H}}{\text{C}} : \overset{\text{H}}{\underset{\text{H}}{\text{C}}} : \right]^{-} [\text{Li}]^{+}
$$

Either way, the result is a dianion that propagates a chain from each end. Other dianionic initiators have been developed.[3,4]

There are a couple of very interesting aspects to these reactions. First, the rates of initiation and propagation vary with the monomer, gegen ion, and solvent. In general, the reactions proceed more rapidly in polar solvents as the species are more highly ionized. (The polarity of the solvent also strongly influences the stereospecific nature of the polymer.) In many important cases the rate of initiation is comparable to the rate of propagation, unlike free-radical addition, in which $r_i \ll r_p$. This means that the initiator (in this case it is an initiator rather than a catalyst) starts chains growing promptly. Second, in the absence of impurities, *there is no termination step*. The chains continue to grow until the monomer supply is exhausted. The ionic chain end is perfectly stable, and the growth of the chains can be resumed by the addition of more monomer. For this reason these materials have been aptly termed *"living" polymers* by Professor Swarc. The presence of proton-donating impurities, such as water or acids, can quickly kill (terminate) them, however.

$$\sim\!\!\!\overset{\displaystyle H}{\underset{\displaystyle H}{C}}\!-\!\overset{\displaystyle H}{\underset{\underset{\bigcirc}{|}}{C:}}\Big]^{-}[Li]^{+} + H_2O \longrightarrow \sim\!\!\!\overset{\displaystyle H}{\underset{\displaystyle H}{C}}\!-\!\overset{\displaystyle H}{\underset{\underset{\bigcirc}{|}}{C:}}H + LiOH$$

This unique mechanism has a number of important practical consequences.

A. Block Copolymerization

If, after the initial monomer charge is exhausted, a second monomer is introduced the chains resume propagation with the second monomer, neatly giving a block copolymer. Monomers can be alternated as desired to give AB, ABA, or other more complicated block structures, conceivably even including three or more monomers.

B. Synthetic Flexibility

Anionic polymerization allows the synthesis of all sorts of interesting and useful molecules. For example, the addition of carbon dioxide to a batch of living chains produces a carboxyl-terminated polymer:

$$\sim\!\!\!\overset{\displaystyle H}{\underset{\displaystyle H}{C:}}\Big]^{-}[Li]^{+} + CO_2 \longrightarrow \sim\!\!\!\overset{\displaystyle H}{\underset{\displaystyle H}{C}}\!-\!\overset{\displaystyle O}{\underset{}{\overset{\|}{C}}}\!-\!O:\Big]^{-}[Li]^{+}$$

$$\sim\!\!\!\overset{\displaystyle H}{\underset{\displaystyle H}{C}}\!-\!\overset{\displaystyle O}{\underset{}{\overset{\|}{C}}}\!-\!O:\Big]^{-}[Li]^{+} + H_2O \longrightarrow \sim\!\!\!\overset{\displaystyle H}{\underset{\displaystyle H}{C}}\!-\!\overset{\displaystyle O}{\underset{}{\overset{\|}{C}}}\!-\!OH + LiOH$$

Similarly, ethylene oxide gives hydroxyl-terminated chains:

$$\sim\!\!\!\overset{\displaystyle H}{\underset{\displaystyle H}{C:}}\Big]^{-}[Li]^{+} + H_2C\!\!-\!\!CH_2 \longrightarrow \sim\!\!\!\overset{H}{\underset{H}{C}}\!-\!\overset{H}{\underset{H}{C}}\!-\!\overset{H}{\underset{H}{C}}\!-\!O:\Big]^{-}[Li]^{+}$$

(with epoxide O below H₂C—CH₂)

$$\sim\!\!\!\overset{H}{\underset{H}{C}}\!-\!\overset{H}{\underset{H}{C}}\!-\!\overset{H}{\underset{H}{C}}\!-\!O:\Big]^{-}[Li]^{+} + H_2O \longrightarrow \sim\!\!\!\overset{H}{\underset{H}{C}}\!-\!\overset{H}{\underset{H}{C}}\!-\!\overset{H}{\underset{H}{C}}\!-\!OH + LiOH$$

Note that if these reagents are added to a lithium-initiated dianion chain, *both* ends of the chain will be capped with the functional group. Such chains are macrodiacids or diols. Carboxyl- or hydroxyl-terminated chains may then take part in the usual condensation reactions.

When monomer is exhausted, if a tetrafunctional monomer such as divinyl benzene (DVB), $H_2C=\overset{H}{\underset{}{C}}\langle O \rangle\overset{H}{\underset{}{C}}=CH_2$, is added to a batch of living chains, it couples to itself and to the living chains. If it is assumed that a DVB can only react with another DVB at one end (there are probably good steric reasons why this should be so), the following type of *star* structures result, with linear branches radiating from a DVB core:

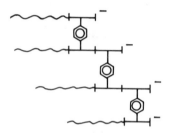

It is conceivable that at this point a second difunctional monomer could be added, giving a star polymer with two different kinds of branches, but, normally, the reactor would be opened and the reaction terminated. From the structure above, there are $(b-1)$ moles of DVB per mole of star branches, where b is the average number of branches per star. Since the moles of star molecules is equal to the moles of living chains over the average number of branches, and with an initiator such as n-BuLi each molecule starts one living chain, to make a b-branch star polymer, one must add

$$\text{moles DVB} = \frac{(b-1)I_0}{b} \tag{11.1}$$

where I_0 = moles initiator charged

Living chains can also be linked by dichloro compounds, for example, instantly doubling their chain lengths:

$$2 \; \sim\overset{H}{\underset{H}{C}}\text{:}\Big]^- [\text{Li}]^+ + \text{Cl}-R-\text{Cl} \longrightarrow \sim\overset{H}{\underset{H}{C}}\text{:}R\text{:}\overset{H}{\underset{H}{C}}\sim + 2\,\text{LiCl}$$

With multifunctional linking agents, this technique can be used to form star polymers with various numbers of branches. It would be nice if all the functional groups on such compounds took part in the linking reaction, but sometimes steric factors lower linking efficiency. Multifunctional initiators can also be used to grow star polymers. Various initiators, linking agents and other reagents for anionic polymerization have been reviewed.[3,4]

C.   Monodisperse Polymers

With $r_i \cong r_p$ and no termination, all growing chains are started nearly simultaneously and compete on an even basis for the available monomer; so all of the chains will be of nearly the same length. As with other modes of polymerization, the number-average chain length is given by the moles of monomer polymerized over the moles of growing chains. With an initiator like n-butyllithium, each molecule starts a single chain; therefore

$$\bar{x}_n = \frac{M_0 - M}{I_0} = \frac{M_0 X}{I_0} \tag{11.2}$$

(If the volume is constant, the moles may be bracketed to give concentrations.) Statistically, the distribution of chain lengths is obtained by answering the question: Given $M_0 - M$ marbles and I_0 buckets, what will be the distribution of marbles among the buckets if they are thrown completely randomly into the buckets? The result is the Poisson distribution

$$\left(\frac{n_x}{N}\right) = \frac{e^{-\bar{x}_n}(\bar{x}_n)^x}{x!} \cong \frac{e^{(x-\bar{x}_n)}}{\sqrt{2\pi x}}\left(\frac{\bar{x}_n}{x}\right)^x \tag{11.3}$$

(as written above, the distribution *does not* include the initiator residue). To a good approximation, the polydispersity index of this distribution is*

$$\frac{\bar{x}_w}{\bar{x}_n} \cong 1 + \frac{1}{\bar{x}_n} \tag{11.4}$$

from which it is seen that even at moderate $\bar{x}_n$'s, the polymer is essentially monodisperse. Figure 11.1 compares the Poisson distribution with the most probable distribution, (9.3) and (10.42), for $\bar{x}_n$'s of 100. Because of their sharpness, the number and weight distributions are almost identical. Not only is anionic polymerization the only known method of synthesizing essentially monodisperse homopolymers, but the blocks in the block copolymers formed this way can also be monodisperse.

Example 1.   Starting with a batch reactor containing 1×10^{-3} g moles of n-butyllithium in dilute solution, suggest *two* methods for making the block copolymer $[S\text{+}_{200}\text{+}B\text{+}_{1000}\text{+}S]_{200}$, where S is the styrene repeating unit and B the butadiene repeating unit.

Solution.   If we let all reactions go to completion, $X = 1$, then from (11.2),

---

*There are the important practical limitations of not being able to add and mix reagents instantaneously and uniformly, and the almost inevitable presence of some terminating impurities. In practice, these always cause some additional spread in the distribution.

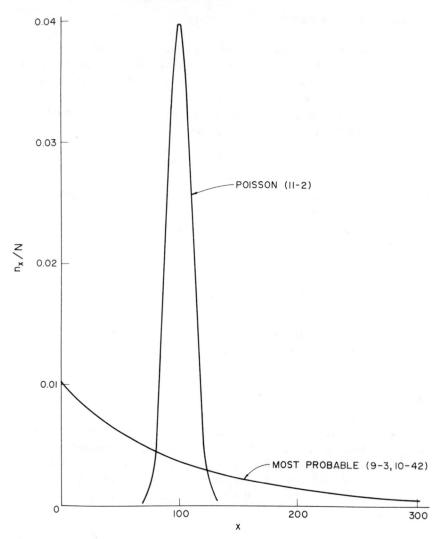

Figure 11.1 Comparison of Poisson and most-probable number-fraction distributions of chain lengths for $\bar{x}_n = 100$.

$M_0 = I_0 x$ (since the blocks will be essentially monodisperse, the bar and n have been left off x).

a.   Add $200 \times 10^{-3} = 0.2$ mole of styrene to the reactor. When it is completely reacted, living polystyrene chains of $x = 200$ will have been formed. Then add $1000 \times 10^{-3} = 1.0$ mole of butadiene, and when it is completely reacted, add another 0.2 mole of styrene and react to completion before opening the reactor and terminating the chains.

b. As before, start with 0.2 mole of styrene. When it is completely reacted, add 0.5 mole of butadiene and react to completion. Then, *link* the resulting 1×10^{-3} moles of $[S]_{200}[B]^-_{500}$ living chains with 0.5×10^{-3} mole of Cl–R–Cl to give 0.5×10^{-3} mole of $[S]_{200}[B]_{500}R[B]_{500}[S]_{200}$. Normally the single R in the middle is insignificant.

Commercially, procedure (b) is preferred. In order to obtain the desired rubbery properties in the central polybutadiene block, nonpolar hydrocarbon solvents are used to promote 1,4 addition. These solvents slow down the crossover reaction from butadiene back to styrene in procedure (a), leading to increased polydispersity in the second polystyrene block. The linking reaction is imperfect, and commercial materials contain some diblock chains along with the desired triblock. Keep in mind also, that each time material is added to a reactor, the probability of some terminating impurities getting in is increased.

It will be left as an exercise for the reader to show how the polymer in Example 1 could be made by starting with a dianionic initiator.

11.3 ANIONIC KINETICS[3,5,6]

A general description of the kinetics of anionic polymerization is complicated by the associations that occur, particularly in nonpolar (hydrocarbon) solvents. The rate of propagation will be proportional to the product of the monomer concentration and the concentration of active living chains $[BuP^-_x Li^+]$

$$r_p = -\left(\frac{d[M]}{dt}\right) = k_p[M][BuP^-_x Li^+] \tag{11.5}$$

Assuming negligible association (as in tetrahydrofuran solvent, for example, or hydrocarbons at BuLi concentrations less than 10^{-4} molar), each initiator molecule will start a growing chain, and in the absence of terminating impurities, the number of active living chains will always equal the number of initiator molecules added:

$$[BuP^-_x Li^+] = [I_0] \tag{11.6}$$

$$r_p = -\left(\frac{d[M]}{dt}\right) = k_p[M][I_0] \tag{11.7}$$

As in the case of condensation polymerization, the chain length increases continuously with conversion and is a function of conversion only in a homogeneous reaction mass. The reaction mass consists only of monomer molecules and polymer chains of essentially a single length. Equation (11.3) characterizes the polymer present.

Example 2.   Consider the anionic batch polymerization of styrene in tetra-hydrofuran solution, $[M_0] = 1.0$ g mole/liter, with *n*-butyllithium initiator. Assuming that $r_i \cong r_p$, that mixing is perfect and instantaneous, and that the volume of the reaction mass is constant,

a.   for an isothermal reaction (constant k_p) obtain an expression relating $\bar{x}_n$ to time.

b.   The reaction is started with $[I_0] = 1 \times 10^{-3}$ g mole/liter. At 50% conversion, water is added instantaneously to a concentration of $[H_2O] = 0.5 \times 10^{-3}$ g mole/liter. At 100% conversion, what chain lengths will be present in the reaction mass?

c.   For case (b), calculate $\bar{x}_n$ at 100% conversion.

Solution

a.   Assume no association in this polar solvent. Separating variables and integrating (11.7) ($[I_0]$ is constant) gives

$$[M] = [M_0]e^{-k_p[I_0]t} \tag{11.8}$$

Plugging into (11.2),

$$x = \frac{[M_0] - [M_0]e^{-k_p[I_0]t}}{[I_0]} = \frac{[M_0]}{[I_0]}(1 - e^{-k_p[I_0]t}) \tag{11.9}$$

(As the polymer will be essentially monodisperse, no bar and subscript are required on x.)

b.   The initiator starts 1×10^{-3} g mole/liter of chains growing. At 50% conversion, the water terminates half the chains. Thus, $\frac{1.0}{4}$ g mole/liter of monomer is present in the 0.5×10^{-3} g mole/liter of terminated chains

$$x \text{ (terminated chains)} = \frac{1.0}{4(0.5 \times 10^{-3})} = 5 \times 10^2$$

The remaining $1.0(\frac{3}{4})$ g mole/liter of monomer continues to grow in 0.5×10^{-3} g mole/liter of unterminated chains.

$$x \text{ (unterminated chains)} = 1.0\left(\frac{3}{4}\right)\frac{1}{0.5 \times 10^{-3}} = 15 \times 10^2$$

c.   Since each initiator molecule starts a chain, at 100% conversion

$$\bar{x}_n = \frac{[M_0]X}{[I_0]} = \frac{1.0}{1 \times 10^{-3}} = 10 \times 10^2$$

or

$$\bar{x}_n = \frac{\Sigma n_x x}{\Sigma n_x} =$$

$$\frac{(0.5 \times 10^{-3})(5 \times 10^2) + (0.5 \times 10^{-3})(15 \times 10^2)}{0.5 \times 10^{-3} + 0.5 \times 10^{-3}} = 10 \times 10^2$$

In Bu Li polymerizations at high concentrations in nonpolar solvents, the chain ends are present almost exclusively as inactive dimers, which dissociate *slightly* according to the equilibrium

$$(BuP_x^- Li^+)_2 \overset{K}{\rightleftharpoons} 2 BuP_x^- Li^+$$

Inactive dimer   Active chains

where

$$K = \frac{[BuP_x^- Li^+]^2}{[(BuP_x^- Li^+)_2]} \ll 1 \qquad (11.10)$$

The concentration of *active chains* is then

$$[BuP_x^- Li^+] = K^{1/2} [(BuP_x^- Li^+)_2]^{1/2} \qquad (11.11)$$

Now it takes two initiator molecules to make one inactive dimer; so

$$\underbrace{\tfrac{1}{2}[BuP_x^- Li^+]}_{\text{negligible}} + [(BuP_x^- Li^+)_2] = \frac{[BuLi]}{2} = \frac{[I_0]}{2} \qquad (11.12)$$

The rate of polymerization then becomes

$$r_p = -\left(\frac{d[M]}{dt}\right) = k_p K^{1/2} \left(\frac{[I_0]}{2}\right)^{1/2} [M] \qquad (11.13)$$

The low value of K, reflecting the presence of most chain ends in the inactive associated state, gives rise to low rates of polymerization in nonpolar solvents. At very high concentrations, association may be even greater, and the rate essentially independent of $[I_0]$.

Example 3.   Which type of isothermal reactor will produce the narrowest possible distribution of chain lengths in an anionic polymerization – batch, continuous stirred tank (backmix) (assume both perfectly stirred), plug-flow tubular, or laminar-flow tubular?

Solution.   As indicated by (11.9) in Example 2, the chain length depends only on how long a chain is allowed to grow. In a batch reactor and an *ideal* plug-flow reactor, all chains react for the same length of time; hence the product will be essentially monodisperse. In a CSTR and a laminar-flow tubular reactor, the residence time of chains in the reactor varies, causing a spread in the distribution. Keep in mind, however, that ideal plug flow is a practical impossibility, particularly with highly viscous polymer solutions. Compare these conclusions with Chapter 10, Example 13.

11.4   HETEROGENEOUS STEREOSPECIFIC POLYMERIZATION[7-10]

The investigations of Ziegler and Natta showed that certain combinations of metal alkyls and metal halides effectively cause the stereospecific polymerization of a variety of monomers by means of an addition mechanism. One example of a *Ziegler–Natta catalyst* system is aluminum triethyl and titanium tetrachloride.

$$Al(C_2H_5)_3 + TiCl_4 \rightarrow \text{complex ppt}$$

Vanadium and cobalt chlorides are also used, as is $Al(C_2H_5)_2Cl$. When these substances are mixed in an inert solvent, a crystalline precipitate is obtained, along with a highly colored supernatant (deep violet or brown). It is known that the reaction involves the reduction of the titanium to a lower valence state, probably $+2$, since $TiCl_3$ will also form an effective catalyst. The supernatant liquid alone will polymerize olefins (aliphatic hydrocarbon monomers, e.g., ethylene, propylene), but the resulting polymers show little stereospecificity. The stereospecific polymerization is thought to occur at active "sites" on the surface of the precipitate formed by reaction of the metal alkyl with metal chloride on the surface of the metal chloride crystals. One postulated mechanism is shown in Fig. 11.2[8] where monomer is chemisorbed at the site (thus accounting for its specific orientation when added to the chain), and the growing chain "unreels" from the site. Hydrogen is used as a chain-transfer agent in these reactions. Chain transfer with the metal alkyl has also been identified:

$$\text{site-chain} + Al(R)_3 \rightarrow \text{site-R} + Al(R)_2\text{-chain}$$

Despite the fact that billions of pounds per year of polymer are produced with them, little is definitely known about the kinetics and mechanisms in Ziegler–Natta catalysis. The results seem to depend on the specific monomer,

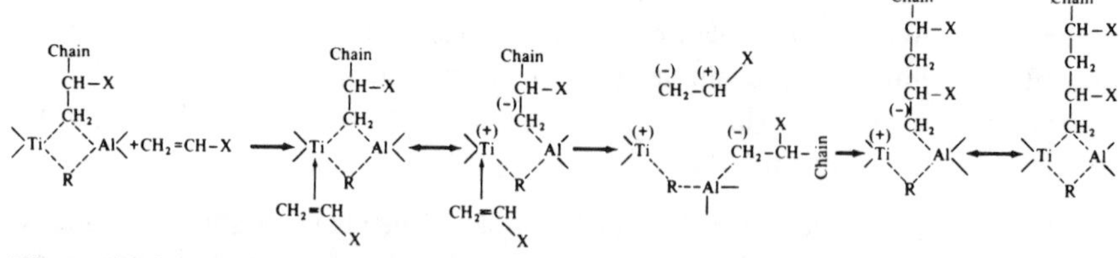

Figure 11.2   Postulated mechanism for Ziegler–Natta catalysis (8). Excerpted by special permission from CHEMICAL ENGINEERING, April 2, 1962. Copyright 1962, by McGraw-Hill, Inc., New York, N.Y. 10036.

the ratio of the catalyst components, and whether the polymer remains in solution as it is formed or precipitates on and coats the catalyst particles. Basically, two approaches are taken to quantify the kinetics. One neglects diffusion external to the catalyst particle and derives Langmuir–Hinshelwood expressions based on the competitive chemisorption of monomer and metal alkyl and on the propagation reaction between chemisorbed monomer and a growing chain on the active site.[11] The other assumes a simple first-order surface reaction and treats the diffusion of monomer to the catalytic surface through the fluid phase and a layer of deposited polymer.[12-14] The former may be more realistic for polymerizations in which the polymer remains in solution, and the latter for gas-phase processes and those in which the polymer precipitates out on the catalyst particles. Regardless, both approaches give rise to complex rate expressions with lots of parameters, and both are capable of describing qualitatively some observed facts. For example, catalyst activity drops off with time. Is this due to chemical deactivation of sites or to increased resistance to monomer diffusion caused by polymer buildup on the catalyst particles? The products of these reactions often have polydispersity indexes of 20 or more. Is this a result of a range of site activities or of variations in the length of the diffusion path of monomer to reach different sites within the polymer-coated catalyst particle? The jury is still out.

Theory aside, Ziegler–Natta catalyst systems are difficult to work with. Great care must be exercised in their preparation and use, as they are easily poisoned and are pyrophoric (burst into flame spontaneously on contact with oxygen). It is a tribute to the chemical industry that they are handled routinely and safely on such a large scale.

Another type of stereospecific polymerization catalyst was developed by the Phillips Petroleum Company. These "Phillips" catalysts are CrO_3 supported on silica or alumina. A postulated mechanism for these catalysts is sketched in Fig. 11.3.

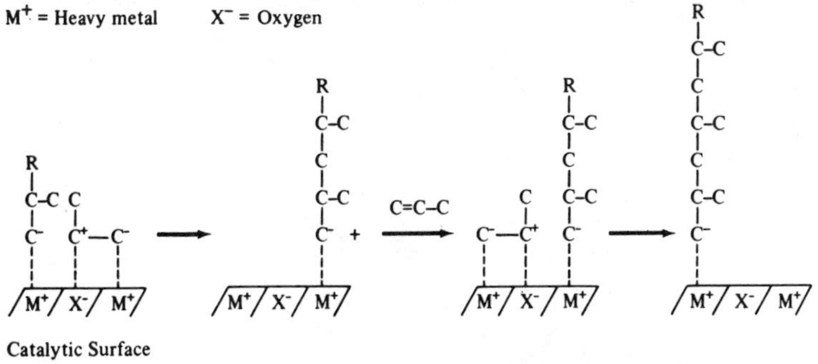

Figure 11.3 Postulated mechanism for Phillips catalysis.[8] Excerpted by special permission from CHEMICAL ENGINEERING, April 2, 1962. Copyright 1962, by McGraw-Hill, Inc., New York, N.Y. 10036.

Not only do the Ziegler—Natta and Phillips catalysts result in the stereo-specific polymerization of unsymmetrical monomers, they are also capable of producing relatively linear polyethylene (the "Phillips" catalysts produces the most linear of all) at low pressures. The older free-radical ethylene polymerizations require much higher pressures and give branched products.

REFERENCES

1. P. H. Plesch, Ed., *The Chemistry of Cationic Polymerization*, MacMillan, New York, 1963.
2. K. E. Russell and G. J. Wilson, "Cationic Polymerizations," *Polymerization Processes*, C. E. Schildknecht and I. Skeist, Eds., Wiley, New York, 1977, Chapter 10.
3. M. Morton and L. J. Fetters, "Anionic Polymerizations and Block Copolymers," *Polymerization Processes*, Wiley, New York, 1977, Chapter 9.
4. E. Franta and P. Rempp, *Polym. Prepr.*, **20**, 1, 5 (1979).
5. R. W. Lenz, *Organic Chemistry of Synthetic High Polymers*, Interscience, New York, 1967, Chapter 13.
6. G. Odian, *Principles of Polymerization*, McGraw-Hill, New York, 1970, Chapter 5.
7. L. Reich and A. Schindler, *Polymerization by Organometallic Compounds*, Interscience, New York, 1966.
8. E. Guccione, *Chem. Eng.* April 2, 1962, p. 93.
9. T. Keii, *Kinetics of Ziegler-Natta Polymerization*, Chapman and Hall, London, 1972.
10. J. C. W. Chien, Ed., *Coordination Polymerization*, Academic Press, New York, 1975.
11. P. J. T. Tait, *Chem. Technol.*, **5**, 688 (1975).
12. V. W. Buls and T. L. Higgins, *J. Polym. Sci. A-1*, **8**, 1025, 1037 (1970); *J. Polym. Sci., Polym. Chem. Ed.*, **11**, 925 (1973).
13. W. R. Schmeal and J. R. Street, *Am. Inst. Chem. Eng. J.*, **17**, 1188 (1971); *J. Polym. Sci., Polym. Phys. Ed.*, **10**, 2173 (1972).
14. N. F. Brockmeier, *Polymerization Reactors and Processes*, J. N. Henderson and T. C. Bouton, Eds., American Chemical Society, Washington, 1979, Chapter 9.

CHAPTER 12

Copolymerization

12.1 MECHANISM

A quantitative treatment of copolymerization is based on the assumption that the reactivity of a growing chain depends only on its active terminal unit. Therefore, when two monomers, M_1 and M_2, are copolymerized, there are four possible propagation reactions:*

Reaction	Rate Equation	
$P_1 \cdot + M_1 \xrightarrow{k_{11}} P_1 \cdot$	$k_{11}[P_1 \cdot][M_1]$	(12.1)
$P_1 \cdot + M_2 \xrightarrow{k_{12}} P_2 \cdot$	$k_{12}[P_1 \cdot][M_2]$	(12.2)
$P_2 \cdot + M_2 \xrightarrow{k_{22}} P_2 \cdot$	$k_{22}[P_2 \cdot][M_2]$	(12.3)
$P_2 \cdot + M_1 \xrightarrow{k_{21}} P_1 \cdot$	$k_{21}[P_2 \cdot][M_1]$	(12.4)

Application of the steady-state assumption to the radicals $P_1 \cdot$ and $P_2 \cdot$ requires that they must be generated and consumed at equal rates.

$P_1 \cdot$'s are generated in reaction (12.4) and consumed in reaction (12.2). Note that (12.1) just converts one $P_1 \cdot$ into another one, with no net change in their number. Therefore

$$k_{12}[P_1 \cdot][M_2] = k_{21}[P_2 \cdot][M_1] \qquad (12.5)$$

*The notation has been changed a bit here to conform to the existing literature. The subscripts 1 and 2 designate the two monomers being copolymerized, *not* the number of repeating units. Thus $P_1 \cdot$ represents a growing chain radical with a terminal unit of monomer 1, that is, a chain radical to which a monomer M_1 was the last added. While mechanisms for copolymerization will be illustrated for free-radical addition, the same general sort of thing occurs in ionic copolymerizations with different quantitative parameters.

The rates of consumption of monomers M_1 and M_2 are

$$\frac{-d[M_1]}{dt} = k_{11}[P_1 \cdot][M_1] + k_{21}[P_2 \cdot][M_1] \tag{12.6}$$

$$\frac{-d[M_2]}{dt} = k_{12}[P_1 \cdot][M_2] + k_{22}[P_2 \cdot][M_2] \tag{12.7}$$

Dividing (12.6) by (12.7) and eliminating the $[P \cdot]$'s with (12.5) gives

$$\frac{d[M_1]}{d[M_2]} = \frac{[M_1]}{[M_2]}\left(\frac{r_1[M_1] + [M_2]}{[M_1] + r_2[M_2]}\right) \tag{12.8}$$

where r_1 and r_2 = *reactivity ratios*

$r_1 = k_{11}/k_{12}$ = relative preference of $[P_1 \cdot]$ for M_1/M_2
$r_2 = k_{22}/k_{21}$ = relative preference of $[P_2 \cdot]$ for M_2/M_1

Reactivity ratios are experimentally determined[1] or may be estimated.[2] In organic free-radical copolymerizations, they are pretty much independent of initiator and solvent, and are only weakly temperature dependent. In ionic co-polymerizations, however, they depend strongly on the gegen ion and solvent.

    This relation may be put in a more convenient form by defining

f_1 = mole fraction of monomer 1 in the reaction mass *at any instant*

F_1 = mole fraction of monomer 1 in the polymer formed *at any instant.*

$$F_1 = 1 - F_2 = \frac{d[M_1]}{d([M_1] + [M_2])} \tag{12.9}$$

$$f_1 = 1 - f_2 = \frac{[M_1]}{[M_1] + [M_2]} \tag{12.10}$$

Combining (12.8)–(12.10) gives

$$F_1 = \frac{r_1 f_1^2 + f_1 f_2}{r_1 f_1^2 + 2f_1 f_2 + r_2 f_2^2} = \frac{(r_1 - 1)f_1^2 + f_1}{(r_1 + r_2 - 2)f_1^2 + 2(1 - r_2)f_1 + r_2} \tag{12.11}$$

    The quantity F_1, the *instantaneous copolymer composition*, is analogous to $\bar{x}_n$, the instantaneous number-average chain length. Like $\bar{x}_n$, it depends on the conditions in the reactor at a particular instant. It, too, is really an average, as not all copolymer chains formed at a particular instant have exactly the same composition. However, the instantaneous distribution of chain compositions is normally very much narrower than the instantaneous distribution of chain lengths, and because the fact that it is an average is not normally of great practical significance and can't be controlled anyhow, the overbar is left off F_1.

12.2 SIGNIFICANCE OF REACTIVITY RATIOS

To gain an appreciation for the physical significance of (12.11), let's look at some special cases of the reactivity ratios.

Case 1: $r_1 = r_2 = 0$

With both reactivity ratios zero, neither radical can regenerate itself; hence a *perfectly alternating* copolymer results, $F_1 = 0.5$ always, until one of the monomers is used up, at which point, polymerization stops.

Case 2: $r_1 = r_2 = \infty$

Here $P_1 \cdot$ radicals can add only M_1 monomer, and $P_2 \cdot$ can add only M_2; so the polymer formed will be a *mixture* of homopolymer 1 and homopolymer 2.

Case 3: $r_1 = r_2 = 1$

Under these conditions, the growing chain radicals can't distinguish between the two monomers, and so the addition depends only on the ratio of monomers in the vicinity of the chain ends; $F_1 = f_1$.

Case 4: $r_1 r_2 = 1$

This is the so-called ideal copolymerization, where each radical displays the same preference for one of the monomers over the other: $k_{11}/k_{12} = k_{21}/k_{22}$; so it doesn't matter what's on the end of the chain. In this case (12.11) reduces to

$$F_1 = \frac{r_1 f_1}{f_1(r_1 - 1) + 1} \tag{12.12}$$

The reader familiar with distillation theory will note here the exact analogy between (12.12) and the vapor–liquid equilibrium composition relation for ideal solutions with a constant relative volatility.

Case 5: $r_1 < 1, r_2 < 1$

This situation corresponds to an azeotrope in vapor–liquid equilibrium. At the azeotropic composition, $F_1 = f_1 = (1 - r_2)/(2 - r_1 - r_2)$.

The case in which r_1 and r_2 are both greater than 1 does not occur.

12.3 VARIATION OF COMPOSITION WITH CONVERSION

In general, $F_1 \neq f_1$; that is, the composition of the polymer formed at any instant will differ from that of the monomer mass. Thus, as the reaction proceeds, the unreacted monomer batch will be depleted in the more reactive monomer, and as the composition of the unreacted monomer changes, so will that of the polymer being formed, according to (12.11).

Example 1.   Draw curves of instantaneous copolymer composition F_1 versus monomer composition f_1 for the following systems, and indicate the direction of composition drift as the reaction proceeds.

a.   Butadiene (monomer 1), styrene (monomer 2), $60°C; r_1 = 1.39, r_2 = 0.78$.
b.   Vinyl acetate (monomer 1), styrene (monomer 2), $60°C; r_1 = 0.01, r_2 = 55$.
c.   Maleic anhydride (monomer 1), isopropenyl acetate (monomer 2), $60°C$;
$r_1 = 0.002, r_2 = 0.032$.

Solution.   Application of (12.11) gives the plots in Fig. 12.1. Note that system (a) approximates ideal copolymerization, case 4 above. In system (b), styrene is the preferred monomer regardless of the terminal radical; hence the copolymer is largely styrene until styrene monomer is nearly used up. System (c) approximates case 1 above. The direction of composition drift with conversion is indicated by

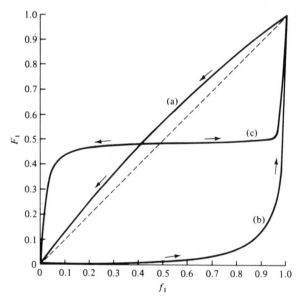

Figure 12.1   Instantaneous copolymer composition (F_1) versus monomer composition (f_1). (a) Butadiene (1), styrene (2), $60°C; r_1 = 1.39, r_2 = 0.78$. (b) Vinyl acetate (monomer 1), styrene (monomer 2), $60°C; r_1 = 0.01, r_2 = 55$. (c) Maleic anhydride (monomer 1), isopropenyl acetate (monomer 2); $r_1 = 0.002, r_2 = 0.032$. The direction of composition drift in a batch reactor is indicated by arrows (Example 1).

arrows. Note that system (c) forms an azeotrope at $F_1 = f_1 = 0.493$. For $f_{1_0} >$ 0.493, F_1 and f_1 will increase with conversion, and vice versa.

Consider a batch consisting of a total of M moles of monomer ($M = M_1 + M_2$). At time t, the monomer has a composition f_1. In the time interval dt, dM moles of monomer polymerize to form dM moles of copolymer with a composition F_1. Therefore, at time $t + dt$, there are left $(M - dM)$ moles of monomer whose composition has been changed to $(f_1 - df_1)$. Writing a material balance on monomer 1, M_1 in monomer at $t = M_1$ in monomer at $(t + dt) + M_1$ in polymer formed in interval dt

$$f_1 M = (M - dM)(f_1 - df_1) + F_1 dM \qquad (12.13)$$

Expanding and neglecting second-order differentials gives

$$\frac{dM}{M} = \frac{df_1}{(F_1 - f_1)} \qquad (12.14)$$

At the start of the reaction, there are present M_0 moles of monomer with a composition f_{1_0}, and at some later time, there are M moles of monomer left with a composition f_1. Integrating between these limits.

$$\ln \frac{M}{M_0} = \int_{f_{1_0}}^{f_1} \frac{df_1}{(F_1 - f_1)} \qquad (12.15)$$

This equation is the exact analog of the Rayleigh equation relating the amount and composition of the still-pot liquid in a batch distillation. Choosing values for f_1 and calculating the corresponding F_1's from (12.11), we may evaluate the integral graphically or numerically, thereby giving a relation between the monomer composition and conversion $(1 - (M/M_0))$. An analytic solution to (12.11) and (12.15) has also been obtained.[3] For $r_1 \neq 1$, $r_2 \neq 1$ (if either is equal to 1, see the original reference),

$$\frac{M}{M_0} = \left(\frac{f_1}{f_{1_0}}\right)^{\alpha} \left(\frac{f_2}{f_{2_0}}\right)^{\beta} \left(\frac{f_{1_0} - \delta}{f_1 - \delta}\right)^{\gamma} \qquad (12.16)$$

where $\alpha = r_2/(1 - r_2)$
$\beta = r_1/(1 - r_1)$
$\gamma = (1 - r_1 r_2)/(1 - r_1)(1 - r_2)$
$\delta = (1 - r_2)/(2 - r_1 - r_2)$

Knowing the monomer composition f_1 as a function of conversion immediately gives the *instantaneous* copolymer composition F_1 as a function of conversion through (12.11). This is important to know because if there is a large variation in the composition of the copolymer chains formed from the beginning of the reaction to high conversions, there may be a wide variation in their *properties* also. For example, if there is a large enough variation in the refractive indices of the materials formed at low and high conversions, the resulting copolymer may be hazy even though completely amorphous.

In addition to the instantaneous copolymer composition F_1, another quantity of interest is $\langle F_1 \rangle$, the *cumulative composition* of the copolymer which has been formed in the batch up to a particular conversion. $\langle F_1 \rangle$ is exactly analogous to $\langle \bar{x}_n \rangle$, the cumulative number-average chain length. It is obtained from a material balance:

moles M_1 charged $=$ moles M_1 in polymer $+$ moles M_1 left in monomer

$$f_{1_0} M_0 \quad = \quad \langle F_1 \rangle (M_0 - M) \quad + \quad f_1 M \qquad (12.17)$$

Rearranging,

$$\langle F_1 \rangle = \frac{f_{1_0} - f_1 (M/M_0)}{(1 - (M/M_0))} \qquad (12.18)$$

The distillation analog of this equation tells the well-educated bootlegger how much of his 20-proof sour mash he must distill over to have 150 proof white lightning in the barrel under his condenser. Figure 12.2 illustrates the quantities defined in terms of the distillation analog.

Example 2. For the styrene–butadiene system of Example 1, sketch curves of instantaneous copolymer composition F_1 and cumulative copolymer composition $\langle F_1 \rangle$ versus conversion, for a batch reaction starting with a 50–50 (mole %) initial monomer charge.

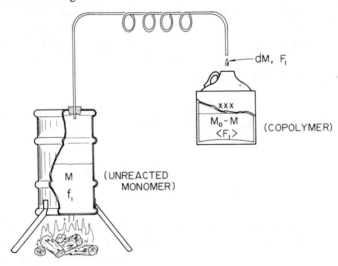

Figure 12.2 Batch distillation analogy to copolymerization. Holdup in vapor space and lines is neglected.

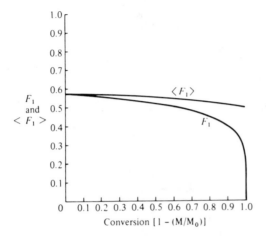

Figure 12.3   Instantaneous (F_1) and cumulative $(\langle F_1 \rangle)$ copolymer composition versus conversion; $r_1 = 1.39$, $r_2 = 0.78$, $f_{10} = 0.50$ (Example 2).

Solution.   Calculations are facilitated by choosing f_1 as the independent variable. As indicated in Example 1, butadiene is the more reactive monomer; so as the reaction proceeds, both polymer and monomer will be enriched in styrene, and f_1 will vary between 0.5 and 0. For a given value of f_1, the conversion, $1 - (M/M_0)$, is calculated from (12.16) with $\alpha = 3.55$, $\beta = -3.57$, $\gamma = 0.983$, and $\delta = -1.29$. For the chosen value of f_1, F_1 is obtained as in Example 1. $\langle F_1 \rangle$ is calculated from (12.18) using the chosen f_1 and the value of M/M_0 obtained from (12.16).

The results are plotted in Fig. 12.3. Note that in this particular case, $\langle F_1 \rangle$ does not vary much between 0% and 100% conversion, but the composition of the chains formed, F_1, changes considerably, particularly at high conversions. Note also that $\langle F_1 \rangle (X = 1.0) = f_{1_0}$ *always.*

Example 3.   Suggest three techniques for producing copolymers of fairly uniform composition, that is, those in which F_1 does not vary much.

Solution.   Three possibilities are sketched in Fig. 12.4. With a semibatch reactor, the more reactive monomer must be replenished as the reaction proceeds to maintain f_1 (and therefore F_1) constant. A method for calculating the appropriate rate of addition has been described.[4] In a continuous stirred-tank (backmix) reactor, both f_1 and F_1 are constant with time. In a continuous plug-flow reactor, the variation in F_1 can be kept small by limiting the conversion per pass in the reactor. Note that the last two techniques require facilities for separating unreacted monomer from the polymer and, in most cases, recycling it.

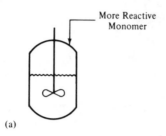

(a)

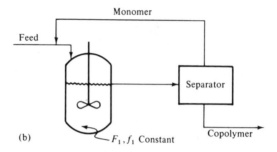

(b)

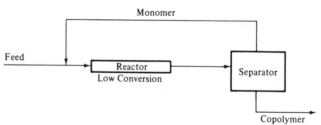

Figure 12.4   Techniques for minimizing the spread of copolymer composition. (a) Semibatch reactor; (b) continuous stirred-tank reactor; (c) tubular reactor.

Example 4.   Consider the semibatch reactor illustrated in Fig. 12.4a.

Let $P(t) =$ moles of monomer (both) in polymer formed to time t
   $A(t) =$ moles of more reactive monomer added to time t

In this setup, if the relation between $A(t)$ and $P(t)$ is arranged properly, not only will the copolymer be of uniform composition, but both monomers will be used up simultaneously. Obtain an expression that relates $A(t)$ to $P(t)$ and system constants.

Solution. First, make a total material balance on monomer: moles initially charged + moles added = moles unreacted + moles in copolymer.

$$M_0 + A(t) = M(t) + P(t)$$

Then, make a material balance on the more reactive monomer 1:

$$f_{1_0} M_0 + A(t) = M(t)f_1 + P(t)F_1$$

By eliminating $M(t)$ between the two, keeping in mind that if F_1 is to be constant, $\langle F_1 \rangle = F_1$ and $f_1 = f_{1_0}$ (constant)

$$A(t) = \frac{(F_1 - f_{1_0})}{(1 - f_{1_0})} P(t)$$

Thus the more reactive monomer is added in direct proportion to the amount of polymer formed (in terms of moles of monomer polymerized). In practice, the real trick is keeping track of $P(t)$. One approach is to make a dynamic heat balance on the reactor. The heat of polymerization liberated is proportional to $P(t)$.

12.4 COPOLYMERIZATION KINETICS

The problem of copolymerization kinetics is not nearly in such good shape. In addition to the four propagation reactions, there are three possible termination reactions $(P_1 \cdot + P_2 \cdot, \ P_1 \cdot + P_1 \cdot, \ P_2 \cdot + P_2 \cdot)$, each with its own rate constant. A general rate equation has been developed,[5] but because of a lack of independent knowledge of the constants involved and the mathematical complexity, it hasn't been used much. An approximate integrated expression including first-order initiator decay has been presented.[6] It involves a single average termination rate constant. This "constant" would not be expected to remain constant as composition varies. Nevertheless, there appears to be some experimental verification of the relation. In certain instances where the compositions don't vary too much (e.g., in the control of a continuous backmix reactor), simplification of (12.6) and (12.7) to

$$\frac{-d[M_1]}{dt} = K_1 [M_1] \tag{12.19}$$

$$\frac{-d[M_2]}{dt} = K_2 [M_2] \tag{12.20}$$

where the "constants" $K_1 = (k_{11} [P_1 \cdot] + k_{21} [P_2 \cdot])$ and $K_2 = (k_{12} [P_1 \cdot] + k_{22} [P_2 \cdot])$ are determined experimentally may prove satisfactory.

12.5   PENULTIMATE EFFECTS AND CHARGE-TRANSFER COMPLEXES

While they are generally satisfactory, there are certain cases in which the equations developed in Section 12.1 do not adequately describe copolymerization. Two approaches have been taken to remedy these deficiencies: invoking penultimate effects and, more recently, postulating the formation of charge-transfer complexes. In the former, the next-to-last monomer unit in a growing chain also exerts an influence on the addition of the next monomer molecule.[7-10] In the latter, a 1–1 complex forms reversibly between electron-donating and electron-accepting comonomers (introducing an equilibrium constant to the analysis). This complex may then polymerize (from either end — introducing four more reactivity ratios) with the uncomplexed monomers.

Complex formation is particularly helpful in explaining free-radical copolymerizations in systems such as styrene–maleic anhydride. This system forms a 1–1 copolymer over most of the range of monomer composition, and the addition of maleic anhydride greatly enhances the rate of polymerization over that of pure styrene, despite the fact that maleic anhydride will not homopolymerize at a noticeable rate. These observations are consistent with the formation of a strong, readily polymerized complex between the monomers. The general equations to describe such copolymerizations have been presented by Seiner and Litt[11] and applied in a number of special cases.[12-15]

REFERENCES

1. L. J. Young, *Polymer Handbook,* 2nd. ed., J. Brandrup and E. H. Immergut, Eds., Wiley, New York, 1975, Chapter 2–6.
2. T. Alfrey, Jr. and C. C. Price, *J. Polym. Sci.,* **2**, 101 (1947).
3. V. E. Meyer and G. G. Lowry, *J. Polym. Sci.,* A3, 2843 (1965).
4. R. J. Hanna, *Ind. Eng. Chem.,* **49**, 208 (1957).
5. E. H. DeButts, *J. Am. Chem. Soc.,* **72**, 411 (1950).
6. K. F. O'Driscoll and R. S. Knorr, *Macromolecules,* **1**, 367 (1968).
7. E. T. Merz, T. Alfrey, and G. Goldfinger, *J. Polym. Sci.,* **1**, 75.
8. W. G. Barb, *J. Polym. Sci.,* **11**, 117 (1953).
9. G. E. Ham, *J. Polym. Sci.,* **54**, 1 (1961).
10. G. E. Ham, *J. Polym. Sci.,* **61**, 9 (1962).
11. J. A. Seiner and M. Litt, *Macromolecules,* **4**, 308 (1971).
12. M. Litt, *Macromolecules,* **4**, 312 (1971).
13. M. Litt and J. A. Seiner, *Macromolecules,* **4**, 314 (1971).
14. M. Litt and J. A. Seiner, *Macromolecules,* **4**, 316 (1971).
15. H. G. Spencer, *J. Polym. Sci., Polym. Chem. Ed.,* **13**, 1253 (1975).

CHAPTER 13

Polymerization Practice

13.1 BULK POLYMERIZATION

The simplest and most direct method of converting monomer to polymer is known as *bulk* or *mass* polymerization. A typical charge might consist of monomer, a monomer-soluble initiator, and perhaps a chain-transfer agent.

As simple as this sounds, there are some serious difficulties that can be encountered, particularly in free-radical bulk polymerizations. One of them is illustrated in Fig. 13.1,[1] which indicates the course of polymerization for various concentrations of methyl methacrylate (Lucite, Plexiglas, Perspex) in benzene, an inert solvent. The reactions were carefully maintained at constant temperature. At the lower monomer concentrations, the conversion versus time curves are well described by (10.15). At the higher concentrations, however, a distinct acceleration of the rate of polymerization is observed, which does not conform to the classical kinetic scheme. This phenomenon is known variously as *autoacceleration,* the *gel effect,* or the *Tromsdorff* effect.

The reasons for this behavior lie in the difference between the propagation reaction (10.3) and the termination steps (10.4a and b), and the extremely high viscosities of concentrated polymer solutions (10^6 poise might be a ballpark figure). The propagation reaction involves the approach of a small monomer molecule and a growing chain end, whereas termination requires that the ends of two growing chains get together. At high polymer concentrations it becomes exceedingly difficult for the growing chain ends to drag their chains through the entangled mass of dead polymer chains. It is nowhere near as difficult for a monomer molecule to pass through the reaction mass. Thus, the rate of the termination reaction is limited not by the nature of the chemical reaction, but by the rate at which the reactants can diffuse together to react, that is, it is *diffusion controlled.* This lowers the effective rate constant k_t, and since it appears in the denominator of (10.12), the net effect is to increase the rate of

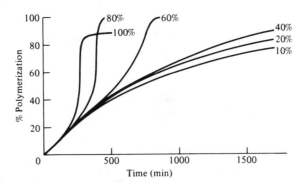

Figure 13.1 Polymerization of methyl methacrylate at $50°C$ in the presence of benzoyl peroxide at various concentrations of monomer in benzene.[1]

polymerization. At very high polymer concentrations and below the temperature at which the chains become essentially immobile — T_g of the monomer-plasticized polymer — even the propagation reaction is diffusion limited, and hence the leveling off of the 100% curve.

The difficulties are compounded by the inherent nature of the reaction mass. Vinyl monomers have rather large exothermic heats of polymerization, typically between -10 and -21 kcal/g mole. Organic systems also have low heat capacities and thermal conductivities, about half those of aqueous systems. To top it all off, the tremendous viscosities prevent effective convective (mixing) heat transfer. As a result, overall heat-transfer coefficients on the order of 1 Btu/hr ft^2 °F are not uncommon, making it difficult to remove the heat generated by the reaction. This raises the temperature, further increasing the rate of reaction (Chapter 10, Example 3) and heat evolution, and can ultimately lead to disaster. To quote Schildknecht[2] on laboratory bulk polymerizations, "if a complete rapid polymerization of a reactive monomer in large bulk is attempted, it may lead to loss of the apparatus, the polymer or even the experimenter."

Example 1. The *maximum possible* temperature rise in a polymerizing batch may be calculated by assuming that no heat is transferred from the system, that is, the adiabatic temperature rise. Estimate the adiabatic temperature rise for the bulk polymerization of styrene, $\Delta H_p = -16.4$ kcal/g mole, molecular weight = 104.

Solution. The polymerization of 1 g mole of styrene liberates 16,400 cal (assuming complete conversion). In the absence of heat transfer, all this energy heats up the reaction mass. The heat capacities of organic compounds are often difficult to find, and since the reaction mass is going from monomer to polymer, which in general have different heat capacities, the heat capacity of the reaction mass changes with conversion, and probably also with temperature. To a reasonable approximation, however, the heat capacity of most liquid organic systems may be taken as 0.5 cal/g $^\circ$C. Thus

$$\Delta T_{max} = \left(16,400\frac{cal}{g\ mole}\right)\left(\frac{1\ g\ mole}{104\ g}\right)\left(\frac{g\ ^\circ C}{0.5\ cal}\right) \cong 315^\circ C(!)$$

Keep in mind that the normal boiling point of styrene is 146°C.

These problems are circumvented in a number of ways:

a. By keeping at least one dimension of the reaction mass small, permitting heat to be conducted out. Polymethylmethacrylate sheets are cast between glass plates at a maximum thickness of an inch or so.

b. By maintaining very low reaction rates through low temperatures and initiator concentrations. The polymerization times for the sheets in (a) are on the order of 30 to 100 h, and the temperatures are raised slowly as the monomer concentration drops. This approach has obvious economic disadvantages.

c. By starting with a *sirup* instead of the pure monomer. A sirup is simply a solution of the polymer in the monomer. It can be made in either of two ways, (1) by carrying the monomer to partial conversion in a kettle, or (2) by dissolving preformed polymer in monomer. Starting off with a sirup means that some of the conversion has already been accomplished, cutting heat generation and monomer concentration in the final polymerization. Since the density of a reaction mass increases on the order of 10–20% between 0% and 100% conversion in a polymerization, the use of a sirup has the added advantage of cutting shrinkage when casting a polymer.

d. By carrying out the reaction continuously, with a lot of heat-transfer surface per unit conversion.

Batch polymerization is generally used to obtain objects with desired shape by polymerizing in a mold. Examples are casting, potting, and encapsulation of electrical components and impregnation of reinforcing agents followed by polymerization. It is also used extensively for the production of thermosetting resins, which are carried to a conversion short of the gel point in the reactor. The crosslinking is later completed in a mold.

Continuous bulk polymerization is becoming increasingly important in the production of thermoplastic molding compounds. A continuous bulk process is outlined in Fig. 13.2.[3] Conversion is carried to about 40% in a stirred tank. The reaction mass then passes down a tower with the temperature increasing to keep the viscosity to a manageable level and to drive up the conversion. The tower may be a simple gravity-flow device, or it may contain slowly rotating spiral

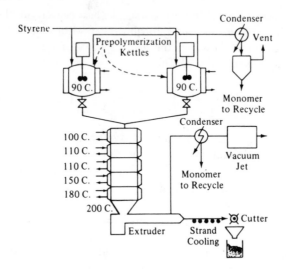

Figure 13.2   Continuous bulk polymerization of styrene. Reprinted by special permission from CHEMICAL ENGINEERING, August 1, 1962. Copyright (C) 1962, by McGraw-Hill, Inc., New York, N.Y. 10036.

blades that scrape the walls, promoting heat transfer and conveying the reaction mass downward. The reaction mass is fed from the tower to a vented extruder at better than 95% conversion. Some additional conversion takes place in the extruder, and a vacuum sucks off unreacted monomer, which may be recycled. The extruded strands of molten polymer are then water cooled and chopped to form the roughly $\frac{1}{8} \times \frac{1}{8} \times \frac{1}{8}$-in. pellets that are sold to processors as molding "powder." Sheets are also continuously cast from sirup between polished sheet-metal belts.

The advantages of bulk polymerization are

1.   As only monomer, initiator, and perhaps chain-transfer agents are necessary, the purest possible polymer may be obtained. This can be important in electrical and optical applications.

2.   Objects may be conveniently cast to shape. If the polymer is one that is crosslinked in the synthesis reaction, this is the *only* way of obtaining such objects short of machining from larger blocks.

3.   Bulk polymerization provides the greatest possible polymer yield per reactor volume.

Among its disadvantages are

1.   It is often difficult to control.

2. To keep it under control, it may have to be run slowly, with the attendant economic disadvantages.

3. As indicated by (10.12) and (10.22), it may be difficult to get both high rates and high average chain lengths because of the opposing effects of [I].

4. It can be difficult to remove the last traces of unreacted monomer. This can be extremely important, for example, if the polymer is intended for use in food-contact applications.

Most bulk polymerizations are homogeneous. However, if the polymer is insoluble in its monomer and precipitates as the reaction proceeds, the process is sometimes known as *heterogeneous bulk* or *precipitation polymerization.* Two examples of such polymers are polyacrylonitrile and polyvinyl chloride (PVC). The latter is produced commercially by a heterogeneous bulk process, which allows control of particle size and porosity for optimum plasticizer absorption. The new gas-phase polyethylene processes[4] might be included in this category, also.

13.2   SOLUTION POLYMERIZATION

The addition of an inert solvent to a bulk polymerization mass minimizes many of the difficulties encountered in bulk systems. As shown in Fig. 13.1, it reduces the tendency toward autoacceleration in free-radical addition. The inert diluent adds its heat capacity without contributing to the evolution of heat, and it cuts the viscosity of the reaction mass at any given conversion. In addition, the heat of polymerization may be conveniently and efficiently removed by refluxing the solvent. Thus, the danger of runaway reactions is minimized.

Example 2.   Estimate the adiabatic temperature rise for the polymerization of a 20% (by weight) solution of styrene in an inert organic solvent.

Solution.   In 100 g of the reaction mass, there are 20 g of styrene; so the energy liberated on its complete conversion to polymer is

$$(20 \text{ g})\left(\frac{1 \text{ g mole}}{104 \text{ g}}\right)\left(\frac{16,400 \text{ cal}}{\text{g mole}}\right) = 3150 \text{ cal}$$

The adiabatic temperature rise is then

$$\Delta T_{max} = (3150 \text{ cal})\left(\frac{\text{g}^\circ\text{C}}{0.5 \text{ cal}}\right)\left(\frac{1}{100 \text{ g}}\right) = 63^\circ\text{C}$$

The advantages of solution polymerization are

1.   Easier heat removal and control.
2.   Since the reactions are more likely to follow known theoretical kinetic relations, the design of reactor systems is facilitated.
3.   For some applications, lacquers, for example, the desired polymer solution is obtained directly from the reactor.

Among its disadvantages are

1.   Since both rate and average chain length are proportional to [M] (in free-radical addition), the use of a solvent lowers them. Additional lowering of $\bar{x}_n$ will occur if the solvent acts as a chain-transfer agent.
2.   Large amounts of expensive, flammable, and perhaps toxic solvent are required.
3.   Separation of the polymer and recovery of the solvent require additional technology.
4.   Removal of the last traces of the solvent and monomer may be difficult (see disadvantage 4 of bulk polymerization in Section 13.1).
5.   Use of an inert solvent in the reaction mass lowers the yield per volume of reactor.

Ionic polymerizations are almost exclusively solution processes. Most Zeigler–Natta polymerizations are solution processes also, although some are run without solvent.[5-8] Figure 13.3 sketches a typical process utilizing a Zeigler–Natta catalyst system. Heat removal from the reactor(s) may be accomplished by refluxing the solvent, using cooling jackets or external pump-around heat

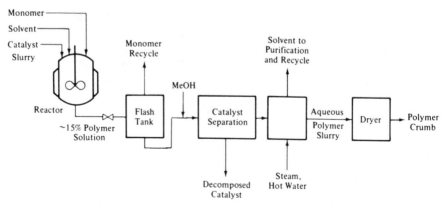

Figure 13.3   Typical Zeigler–Natta polymerization process.

exchangers, or combinations of these. Where a crystalline polymer is being produced, the reaction may be carried out at a temperature low enough such that the polymer precipitates out as it is formed and the product of the reactor is a slurry, rather than a homogeneous solution. The catalyst is normally deactivated with methanol or acid and filtered, centrifuged, or settled out, although much recent emphasis has been placed on improving catalyst yield (grams polymer produced per gram catalyst) so that this difficult and expensive step can be eliminated. Solvent and unreacted monomer are stripped with hot water and steam and recovered, leaving a water slurry of polymer, which is then dried to form a "crumb." With rubbers, the crumb is compacted and baled; with plastics, it is normally extruded and pelletized. Reactor designs for these processes are interesting and extremely varied. A majority of the newer processes are continuous.

13.3   INTERFACIAL POLYCONDENSATION

A variation of solution polymerization known as interfacial polycondensation has been used in the laboratory for a long time,[9] and rumor has it that it is now being applied commercially. One monomer of a condensation pair is dissolved in a liquid, and the other member of the pair in another liquid. The liquids are insoluble in one another. The polymer is soluble in neither and forms at the interface between them. One of the phases generally also contains an agent that reacts with the molecule of condensation to drive the reaction to completion. An example of such a process is the preparation of nylon 6/10 from hexamethylene diamine and sebacoyl chloride (the acid chloride form of sebacic acid).

$$H_2N(CH_2)_6NH_2 + Cl-\overset{O}{\overset{\|}{C}}-(CH_2)_8-\overset{O}{\overset{\|}{C}}-Cl \longrightarrow \left[N-(CH_2)_6-\overset{H}{N}-\overset{O}{\overset{\|}{C}}-(CH_2)_8-\overset{O}{\overset{\|}{C}} \right]_x + HCl$$

The acid chloride is dissolved in CCl_4, for example, and the diamine in water, along with some NaOH to soak up the HCl. In the classic "rope trick" demonstration, the aqueous layer is gently floated on top of the organic layer in a beaker. The reactants diffuse to the interface, where they react very rapidly to form a polymer film. With care, the film can be withdrawn from the interface in the form of a continuous hollow strand which traps considerable liquid. New polymer forms at the interface as the old is withdrawn. Commerically, it is probably easier simply to stir the phases together.

One of the major advantages of this technique is that these reactions usually proceed very rapidly at room temperature and atmospheric pressure, in contrast to the long times, high temperatures, and vacuums usually associated with polycondensations. This must be balanced against the cost of preparing the special monomers, like the acid chloride above, and the need to separate and recycle solvents and unreacted monomers.

13.4   SUSPENSION, BEAD, OR PEARL POLYMERIZATION

In discussing bulk polymerization, it was mentioned that one of the ways of facilitating heat removal was to keep one dimension of the reaction mass small. This is carried to its logical extreme in suspension polymerization by suspending the monomer in the form of droplets 0.01 to 1 mm in diameter in an inert, non-solvent liquid (almost always water). In this way, each droplet becomes a single bulk reactor, but with dimensions small enough so that heat removal is no problem and the heat can easily be soaked up by and removed from the low-viscosity, inert suspension medium.

An important characteristic of these systems is that the suspensions are *thermodynamically unstable* and must be maintained with agitation and suspending agents. A typical charge might consist of

$$
\left.
\begin{array}{l}
\text{monomer (water insoluble)} \\
\text{initiator (monomer soluble)} \\
\text{chain-transfer agent (monomer soluble)}
\end{array}
\right\} \text{monomer phase}
$$

$$
\text{water}
$$

$$
\text{suspending agent} -
\left\{
\begin{array}{l}
\text{protective colloid} \\
\text{insoluble inorganic salt}
\end{array}
\right.
$$

Two types of suspending agent are used. A "protective colloid" is a water-soluble polymer whose function is to increase the viscosity of the continuous water phase. This hydrodynamically hinders coalescence of monomer drops; but it is inert with regard to the polymerization. A finely divided insoluble inorganic salt such as $MgCO_3$ may also be used. It collects at the droplet—water interface by surface tension and prevents coalescence of the drops upon collision. A pH buffer is sometimes also used to help stability.

The monomer phase is suspended in the water at about a $\frac{1}{2}$ to $\frac{1}{4}$ monomer/water volume ratio. The reactor is purged with nitrogen and heated to start the reaction. Once the reaction is under way, temperature control in the reactor is facilitated by the added heat capacity of the water, the low viscosity of the reaction mass — essentially that of the continuous phase — allowing easy heat removal through a jacket.

The size of the product beads depends on the strength of agitation, as well as on the nature of the monomer and suspending system. Between about 20% and 70% conversion, *agitation is extremely critical.* Below 20% the organic phase is still fluid enough to redisperse, and above 70% the particles are rigid enough to

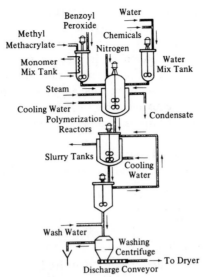

Figure 13.4 Suspension polymerization of methyl methacrylate. Excerpted by special permission from CHEMICAL ENGINEERING, June 6, 1966. Copyright (C) 1966 by McGraw-Hill, Inc., New York, N.Y. 10036.

prevent agglomeration; but if agitation stops or weakens between these limits, the sticky particles will coalesce or agglomerate in a large mass and finish polymerization that way. Again quoting Schildknecht[10] "After such uncontrollable polymerization is completed in an enormous lump, it may be necessary to resort to a compressed air drill or other mining tools to salvage the polymerization equipment."

Since any flow system is bound to have some relatively stagnant corners, it has been impractical to run suspension polymerization continuously. Figure 13.4[11] shows a typical process. The reactors are usually jacketed, stainless or glass-lined steel kettles of up to 30,000-gal capacity. The polymer beads are filtered or centrifuged and either water washed to remove the protective colloid or rinsed with a dilute acid to decompose the $MgCO_3$. The beads are quite easy to handle when wet, but they tend to pick up a static charge when dry, making them cling to each other and everything else. The beads can be molded directly, extruded and chopped to form molding "powder," or used as is, for example, as ion-exchange resins or the beads from which polystyrene foam cups and packing supports are made.

Ion-exchange resins are basically suspension beads of polystyrene crosslinked by copolymerization with a few percent divinyl benzene, which are then treated chemically to provide the necessary functionality. To reduce mass-transfer resistance in the ion-exchange process, an inert solvent may be incorporated in the organic suspension phase. When polymerization is complete, the solvent is removed, leaving a highly porous bead with a large internal surface area (macro-reticular). Foam beads are linear polystyrene containing an inert liquid blowing agent, usually pentane. The pentane may be added to the monomer prior to polymerization, but, more commonly, it is added to the reactor after polymerization and is absorbed by the polystyrene beads. When exposed to steam in a mold, the beads soften and are foamed and expanded by the volatilized blowing agent to form the familiar cups and other foam items.

The major advantages of suspension polymerization, then, are

1. Easy heat removal and control.
2. The polymer is obtained in a convenient, easily handled, and often directly useful form.

Disadvantages include:

1. Low yield per reactor volume.
2. A somewhat less pure polymer than from bulk polymerization, since there are bound to be remnants of the suspending agent(s) absorbed on the particle surface.
3. The inability to run the process continuously, although if several batch reactors are alternated, the process may be continuous from that point on.

13.5   EMULSION POLYMERIZATION

When the supply of natural rubber from the East was cut off by the Japanese in World War II, the United States was left without an essential material. The success of the Rubber Reserve Program in developing a suitable synthetic substitute and the facilities to produce it in the necessary quantities is one of the all-time outstanding accomplishments of chemists and engineers. The styrene—butadiene copolymer rubber GR-S (government rubber—styrene) or SBR (styrene—butadiene rubber) as it is now called (developed during the war) is still the most important synthetic rubber and is still produced, along with a variety of other polymers, largely by the emulsion polymerization process developed then.

The theory behind the emulsion reaction has been discussed in Chapter 10. A typical commercial process is shown in Fig. 13.5. The reactors are usually stainless or glass-lined steel tanks, similar to those used for suspension polymerization. In contrast to suspension polymerization, however, a proper emulsion is thermodynamically stable, and therefore emulsion polymerization can be run continuously. Newer processes often have several continuous stirred-tank reactors in series.

The product of an emulsion polymerization is a *latex*—polymer particles on

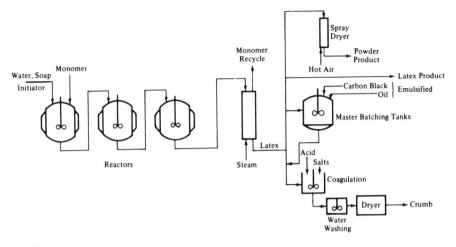

Figure 13.5   Emulsion polymerization process.

the order of 500 to 1500 Å (0.05 to 0.15 μm) stabilized by the soap. These latices are often important items of commerce in their own right. Two familiar examples are "white glue" and latex paints. The latter, also containing pigments and various agents to control application and film-forming characteristics, are water dispersable only as long as the latex is maintained. As soon as the water is evaporated and the water-insoluble polymer particles coalesce to form the paint film, you can no longer clean your brushes with water. Where the polymer must be mixed with other materials, the process of *"master batching"* sometimes allows this to be done conveniently and uniformly. In rubber technology, carbon black and oil are emulsified and mixed with the rubber latex, and then the whole works is coagulated together, giving a uniform and intimate dispersion of the additives in the rubber.

Not too long ago, it appeared that for many polymers, emulsion polymerization was rapidly losing ground to other polymerization processes. However, the increased recognition of possible adverse physiological effects of even small amounts of residual monomer suddenly makes emulsion polymerization look very attractive in certain applications. The extremely small size of the latex particles provides a very short diffusion path for the removal of small molecules from the polymer, by steam stripping, for example, permitting very low levels of residual monomer to be obtained.

For many applications the solid polymer must be recovered from the latex. The simplest method is spray drying, but since no attempt is made to remove the soap, the product is an extremely impure polymer. A latex may be "creamed" by adding a material that is at least a partial solvent for the polymer, acetone, for example. This makes the particles sticky and causes some agglomeration. The latex is then coagulated by adding an acid, sulfuric, for example, which converts the soap to the insoluble hydrogen form, or by adding an electrolyte salt, which disrupts the stabilizing double layer on the particles, causing them to agglomerate through electrostatic attraction. The former method leaves much insoluble material adsorbed on the particle surfaces, but in some applications, this may even be beneficial; for example, the fatty acid acts as a lubricant in tire manufacturing. The coagulated polymer "crumb" is then washed, dried, and either baled or processed further.

In summary, the advantages of emulsion polymerization are

1.   Ease of control: The viscosity of the reaction mass is much less than a true solution of comparable concentration; the water adds its heat capacity; and the reaction mass may be refluxed.

2.   It is possible to obtain both high rates of polymerization and high average chain lengths through the use of high soap and low initiator concentrations.

3.   The latex product is often directly valuable or aids in obtaining uniform compounds through master batching.

4.   The small size of the latex particles allows the attainment of low residual monomer levels.

Its disadvantages are

1.   It is difficult to get pure polymer. The tremendous surface area of the tiny particles provides plenty of room for adsorbed impurities — this includes water attracted by residual soap, traces of which can cause problems in certain applications.

2.   Considerable technology is required to recover the solid polymer.

3.   The water in the reaction mass lowers the yield per reactor volume.

REFERENCES

1. G. V. Shulz and G. Harborth, *Makromol. Chem.*, **1**, 106 (1947).

2. C. E. Schildknecht, *Polymer Processes*, Interscience, New York, 1956, p. 38.

3. M. H. Wohl, *Chem. Eng.*, Aug. 1, 1962, p. 60.

4. R. Martino, *Mod. Plast.*, **55**, 2, 38 (1978).

5. M. Sittig, *Pet. Refiner*, **39**, 11, 162 (1960).

6. M. Sittig, *Pet. Refiner*, **40**, 3, 129 (1961).

7. H. D. Anspon, *Manufacture of Plastics*, W. M. Smith, Ed., Reinhold, New York, 1964, Chapter 2.

8. J. E. Jezl, *Manufacture of Plastics*, W. M. Smith, Ed., Reinhold, New York, 1964, Chapter 3.

9. P. W. Morgan, *Soc. Plast. Eng. J.*, **15**, 6, 485 (1959).

10. C. E. Schildknecht, *Polymer Processes*, Interscience, New York, 1956, p. 94.

11. E. Guccione, *Chem. Eng.*, June 6, 1966, p. 138.

PART 3

Polymer
Properties

CHAPTER 14

Rubber Elasticity

14.1 INTRODUCTION

Natural and synthetic rubbers possess some interesting, unique and useful mechanical properties. No other materials are capable of reversible extension to strains of 6–700%. No other materials exhibit an increase in modulus with increasing temperature. It was recognized long ago that vulcanization was necessary in order that rubber deformation be completely reversible. We now know that this is a result of the crosslinks thereby introduced preventing the bulk slippage of the molecules past one another and eliminating flow (irrecoverable deformation). Thus, when a stress is applied to a sample of crosslinked rubber, equilibrium is established fairly rapidly. Once the rubber is at equilibrium, the properties of the rubber can be described by thermodynamics.

14.2 THERMODYNAMICS OF ELASTICITY

Consider an element of material with dimensions $a \times b \times c$, as sketched in Fig. 14.1. Applying the first law of thermodynamics to this system,

$$dU = dQ - dW \tag{14.1}$$

where dU is the change in the system's *internal energy*, and dQ and dW are the heat and work exchanged between system and surroundings as the system undergoes a differential change. (We have adopted the convention here that work done *by the system on the surroundings* is positive.)

In general, there are three types of mechanical work possible:

1. Work done by a tensile force f:

$$dW \text{ (tensile)} = -f \, dl \tag{14.2}$$

189

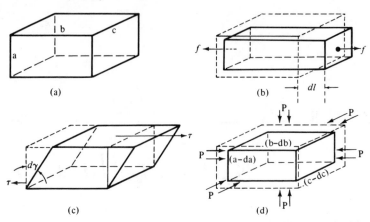

Figure 14.1   Types of mechanical deformation. (*a*) Unstressed; (*b*) pure tension; (*c*) pure shear; (*d*) isotropic compression.

where dl is the differential change in the system's length arising from the application of the force f. This is the fundamental definition of work. The negative sign arises from the need to reconcile the mechanical convention of treating a tensile force (which does work *on* the system) as positive with the thermodynamic convention above.

2.   Work done by a shear stress τ:

$$dW \text{ (shear)} = \text{(force)(distance)} = -(\tau bc)(a\, d\gamma) = -\tau V\, d\gamma \quad (14.3)$$

where γ is the shear strain (Fig. 14.1*c*) and $V = abc =$ the system volume.

3.   Work done by an isotropic pressure in changing the volume:

$$dW \text{ (pressure)} = P(cb)\, da + P(ac)\, db + P(ab)\, dc = P\, dV \quad (14.4)$$

Note that no minus sign is needed here. A positive pressure causes a decrease in volume (negative dV) and does work *on* the system.

If the deformation process is assumed to occur *reversibly* (in a thermodynamic sense).

$$dQ = T\, dS \quad (14.5)$$

where S is the system's entropy.

Combining the preceding five equations gives a general relation for the change of internal energy of an element of material undergoing a differential deformation:

$$dU = T\, dS - P\, dV + f\, dl + V\tau\, d\gamma \quad (14.6)$$

Now, let us consider three individual types of deformation:

1.   Pure tension at constant volume and temperature. Under these con-

ditions, $dV = \tau = 0$. Dividing the remaining terms in (14.6) by dl, restricting to constant T and V, and solving for f gives

$$f = \left(\frac{\partial U}{\partial l}\right)_{T, V} - T\left(\frac{\partial S}{\partial l}\right)_{T, V} \tag{14.7}$$

2. Pure shear at constant volume and temperature. Here, $dV = f = 0$. Dividing the remaining terms in (14.6) by $d\gamma$, restricting to constant T and V, and solving for τ gives

$$\tau = \frac{1}{V}\left(\frac{\partial U}{\partial \gamma}\right)_{T, V} - \frac{T}{V}\left(\frac{\partial S}{\partial \gamma}\right)_{T, V} \tag{14.8}$$

3. Isotropic compression only, at constant temperature:

$$P = -\left(\frac{\partial U}{\partial V}\right)_{T} + T\left(\frac{\partial S}{\partial V}\right)_{T} \tag{14.9}$$

It is very difficult to carry out tensile experiments at constant volume to obtain the partial derivatives in (14.7). Most tests are carried out at constant pressure (atmospheric), and, in general, there is a change in volume with tensile straining. Fortunately, Poisson's ratio is approximately 0.5 for rubbers; so this change in volume is small, and (14.7) is approximately valid for tensile deformation at constant pressure, also. For precise work the hydrostatic pressure must be varied to maintain V constant, or corrections must be applied to the constant-pressure data to obtain the constant-volume coefficients.[1,2] In pure shear, V should be constant and (14.8) valid.

A.   Types of Elasticity

Equations (14.7)–(14.9) reveal that there are *energy* (the first term on the right) and *entropy* (the second term on the right) contributions to the tensile force, shear stress, or isotropic pressure. In polymers, *energy elasticity* represents the storage of energy resulting from the elastic straining of bond angles and lengths like springs from their equilibrium values. *Entropy elasticity* is caused by the decrease in entropy upon straining. This can be visualized by considering a single polymer molecule subjected to a tensile stress. In an unstressed state, it is free to adopt an extremely large number of random "balled-up" configurations, Fig. 14.2 left, switching from one to another through rotation about bond angles. Now imagine it to be stretched out under the application of a tensile force, Fig. 14.2 right. It is obvious that there are far fewer configuration possibilities. The more the molecule is stretched, the fewer configurations there are. Now $S = k \ln \Omega$, where k is Boltzmann's constant and Ω is the number of

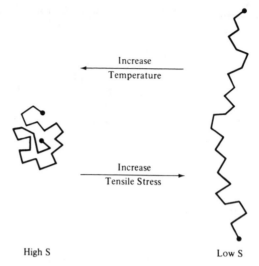

Figure 14.2   The effects of stress and temperature on chain configurations.

configurational possibilities; so *stretching decreases the entropy* (increases the order). Raising the temperature has precisely the opposite effect. The added thermal energy of the chain segments increases the intensity of their lateral vibrations, favoring a return to the more random or higher entropy state. This tends to pull the extended chain ends together, giving rise to a retractive force.

B.   The "Ideal" Rubber

In a gas subjected to an isotropic pressure, the energy term in (14.9) arises from the change in intermolecular forces with volume, and the entropy term from the increased room (and therefore greater "disorder") the molecules gain with volume. In an *ideal gas* there are *no* intermolecular forces, $(\partial U/\partial V)_T = 0$. By analogy, in an *ideal rubber* $(\partial U/\partial l)_{T, V} = (\partial U/\partial \gamma)_{T, V} = 0$, and elasticity arises only from entropy effects. For many gases around room temperature and above, and around atmospheric pressure and below, $(\partial U/\partial V)_T < T(\partial S/\partial V)_T$, and the ideal gas law is a good approximation. Similarly, as illustrated in Fig. 14.3, $(\partial U/\partial l)_{T, V} < T(\partial S/\partial l)_{T, V}$ for rubbers, and they behave approximately as ideal rubbers.

C.   Effects of Temperature at Constant Force

Now let us consider what happens to the length of a piece of rubber when its temperature is changed while a weight is suspended from it, that is, when it is maintained at a constant tensile force. Assuming constant volume (an approxi-

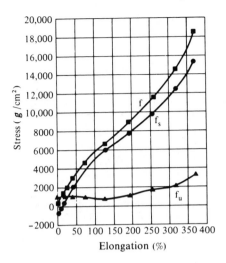

Figure 14.3   Energy (f_U) and entropy (f_S) contributions to tensile stress in natural rubber at 20°C. Reprinted from reference (3). Copyright 1942 by the AMERICAN CHEMICAL SOCIETY. Reprinted by permission of the copyright owner.

mation, in the usual constant-pressure experiment), $dU = T\,dS + f\,dl$. Solving for l, dividing by dT, and restricting to constant f as well as V,

$$\left(\frac{\partial l}{\partial T}\right)_{f,V} = \frac{1}{f}\left(\frac{\partial U}{\partial T}\right)_{f,V} - \frac{T}{f}\left(\frac{\partial S}{\partial T}\right)_{f,V} \tag{14.10}$$

As before, the first term on the right represents energy elasticity and the second, entropy elasticity. Since internal energy generally increases with temperature, the partial derivative in the energy term is positive, as is f. The energy term, therefore, causes an increase in length with temperature, (positive $(\partial l/\partial T)_{f,V}$). This is the normal thermal expansion, observed in all materials: metals, ceramics, glasses, and so on, reflecting the increase in the average distance between atomic centers with temperature. All factors in the entropy term are positive, however, and since it is preceded by the negative sign, it gives rise to a *decrease* in length with increasing temperature. In rubbers, where the entropy effect overwhelms the normal thermal expansion, this is what is actually observed. In all other materials where the structural units are confined to a single arrangement (e.g., the atoms in a crystal lattice cannot readily interchange), the entropy term is negligible. The magnitude of the entropy contraction in rubbers is much greater than the thermal expansion of other materials. An ordinary rubber band will contract an inch or so when heated to 300°F under stress, while the expansion of a piece of metal of similar length over a similar temperature range would not be noticeable to the naked eye.

D.   Effects of Temperature at Constant Length

It is interesting to consider what happens to the force in a piece of rubber when it is heated while stretched to a constant length. Using the exact thermodynamic Maxwell relation

$$-\left(\frac{\partial S}{\partial l}\right)_{T,V} = \left(\frac{\partial f}{\partial T}\right)_{l,V} \tag{14.11a}$$

to describe approximately the usual experiment conducted at constant pressure (a better approximation to the constant pressure experiment is

$$-\left(\frac{\partial S}{\partial l}\right)_{T,V} \cong \left(\frac{\partial f}{\partial T}\right)_{P,\alpha} \tag{14.11b}$$

where α, the extension ratio, is equal to l/l_0, the ratio of stretched to unstretched length at a particular temperature), combining (14.11a) and (14.7) gives

$$\left(\frac{\partial f}{\partial T}\right)_{l,V} = \frac{f}{T} - \frac{1}{T}\left(\frac{\partial U}{\partial l}\right)_{T,V} \tag{14.12}$$

Now both f and T are positive; so the first term on the right causes the force to increase with temperature, a result of the greater thermal agitation (tendency toward higher entropy) of the extended chains. The partial derivative in the second term is also positive, as energy is stored springlike in the strained bond angles and lengths with extension. With the negative sign in front, this term predicts a relaxation of the tensile force with increasing temperature. Again, this second term reflects the ordinary thermal expansion obtained with all materials, but in rubbers, at reasonably large value of f, it is overshadowed by the first (entropy) term, and the force increases with temperature. For an *ideal* rubber, $(\partial U/\partial l)_{T,V} = 0$, and integration of (14.12) at constant volume gives

$$f = (\text{constant})\,T \qquad (\text{ideal rubber}) \tag{14.13}$$

This is analogous to the linearity between P and T in an ideal gas at constant V. These observations are confirmed in Fig. 14.4. The negative slope at low elongations arises from the predominance of thermal expansion when elongation, and hence f, is low. Note that there is an intermediate elongation, the *thermoelastic inversion* point, at which force will be essentially independent of temperature, where thermal expansion and entropy contraction balance.

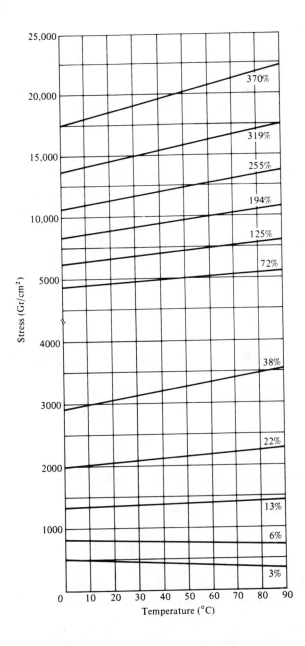

Figure 14.4   Force versus temperature in natural rubber maintained at constant extension (% relative to length at 20 C).[3] (Copyright 1942 by the American Chemical Society. Reprinted by permission of the copyright owner.)

14.3   STATISTICS OF IDEAL RUBBER ELASTICITY[1,4,5]

A typical rubber consists of long chains connected by short crosslinks every few hundred carbon atoms. The chain segments between crosslinks are known as *network chains*. The change in entropy upon stretching a sample containing N moles of network chains is

$$S - S_0 = NR \ln \Omega / \Omega_0 \qquad (14.14)$$

where the subscript 0 refers to the unstretched state, Ω is the number of configurations available to the N moles of network chains, and R is the gas constant. By statistically evaluating the Ω's, it is possible to show that for constant-volume stretching,

$$S - S_0 = -\tfrac{1}{2} NR \left[\left(\frac{l}{l_0} \right)^2 + 2 \left(\frac{l_0}{l} \right) - 3 \right] \qquad (14.15)$$

For an *ideal rubber*, in which the tensile force is given by

$$f = -T \left(\frac{\partial S}{\partial l} \right)_{T, V} \qquad \text{(ideal rubber)} \qquad (14.16)$$

differentiation of (14.15) and insertion into (14.16) then gives

$$f = \frac{NRT}{l_0} \left[\left(\frac{l}{l_0} \right) - \left(\frac{l_0}{l} \right)^2 \right] \qquad (14.17)$$

But,

$$N = \frac{\text{mass}}{\bar{M}_c} = \frac{\rho V}{\bar{M}_c} = \frac{\rho l_0 A_0}{\bar{M}_c} = \frac{\rho l A}{\bar{M}_c} \qquad (14.18)$$

where $\bar{M}_c$ is the number-average molecular weight of the network chains, ρ the density, and A the cross-sectional area of the sample (since the volume change in stretching a piece of rubber is negligible, $A_0 l_0 = Al$). Therefore,

$$f = \frac{\rho A_0 RT}{\bar{M}_c} \left[\left(\frac{l}{l_0} \right) - \left(\frac{l_0}{l} \right)^2 \right] \qquad (14.19)$$

The *engineering tensile stress* is defined as the tensile force divided by the initial cross-sectional area of the sample and is, therefore,

$$\sigma \text{ (engineering)} = \frac{f}{A_0} = \frac{\rho RT}{\bar{M}_c} \left[\left(\frac{l}{l_0} \right) - \left(\frac{l_0}{l} \right)^2 \right] \qquad (14.20)$$

The *true tensile stress*, defined as the tensile force over the actual area A at length l, is

$$\sigma \text{ (true)} = \frac{f}{A} = \frac{\rho RT}{\bar{M}_c} \left[\left(\frac{l}{l_0} \right)^2 - \left(\frac{l_0}{l} \right) \right] \qquad (14.21)$$

Since the tensile strain ϵ is

$$\epsilon = \frac{(l - l_0)}{l_0} \tag{14.22}$$

the slope of the true stress–strain curve (tangent Young's modulus) is

$$E = \left(\frac{\partial \sigma}{\partial \epsilon}\right)_T = \left(\frac{\partial \sigma}{\partial l}\right)_T \left(\frac{\partial l}{\partial \epsilon}\right)_T = \frac{\rho RT}{\bar{M}_c} \left[2\left(\frac{l}{l_0}\right) + \left(\frac{l_0}{l}\right)^2 \right] \tag{14.23}$$

and the initial modulus (as $l \rightarrow l_0$) for an ideal rubber becomes

$$E \,(\text{initial}) = \frac{3\rho RT}{\bar{M}_c} \tag{14.24}$$

Equations (14.19)–(14.24) point out two important facts: first, that the force (or modulus) in an ideal rubber sample held at a particular strain increases in proportion to the absolute temperature, confirming (14.13), and second, that it is *inversely* proportional to the molecular weight of the chain segments between crosslinks. Thus increased crosslinking, which reduces $\bar{M}_c$, is an effective means of stiffening a rubber. Equation (14.24) is often used to obtain $\bar{M}_c$ from mechanical tests and thereby evaluate the efficiency of various crosslinking procedures. Even noncrosslinked polymers exhibit rubbery behavior above their T_g's for short periods of time. This is due to mechanical entanglements acting as temporary crosslinks, and $\bar{M}_c$ then represents the average length of chain segments between entanglements.

When compared with experimental data, (14.19) does a good job in compression but begins to fail at extension ratios (l/l_0) greater than about 1.5, where the experimental force becomes greater than predicted. There are a number of reasons for this. First, (14.15) is based on the assumption of a Gaussian distribution of network chains. This assumption fails at high elongations, and it is also in error if crosslinks are formed when chains are in a strained configuration. Second, it does not take into account the presence of chain end segments, which do not contribute to the support of stress. Third, some rubbers (natural, in particular) may begin to crystallize as a result of chain orientation at high elongations. This causes the stress–strain curve to shoot up markedly. Modifications to the theory are available for the first two factors, improving things considerably in the absence of crystallinity, but there is as yet no satisfactory treatment of the effect of crystallinity on stress–strain properties.

It is important to keep in mind also, that in practice, rubbers are rarely used in the form of pure polymer. They almost always are reinforced with carbon black and often contain other fillers, plasticizing and extending oils, and so forth, all of which influence the stress–strain properties and are not considered by the theories discussed.

REFERENCES

1. P. J. Flory, *Principles of Polymer Chemistry,* Cornell University Press, Ithaca, New York, 1953, Chapter 11.
2. P. Meares, *Polymers: Structure and Bulk Properties,* Van Nostrand, Princeton, N.J., 1965, Chapter 6.
3. R. L. Anthony, R. H. Caston, and E. Guth, *J. Phys. Chem.,* **46,** 826 (1942).
4. A. V. Tobolsky, *Properties and Structure of Polymers,* Wiley, New York, 1960, Chapter 2.
5. Meares, *Polymers: Structure and Bulk Properties,* Van Nostrand, Princeton, N.J., 1965, Chapters 7, 8.

CHAPTER 15

Purely Viscous Flow

15.1 INTRODUCTION

*Rheology** is the science of the deformation and flow of materials. Much contemporary rheological work is concerned directly with polymer systems because they exhibit such interesting, unusual, and difficult-to-describe (at least from the standpoint of traditional materials) deformation behavior. The simple and traditional linear engineering models, Newton's law (for flow) and Hooke's law (for elasticity), often just are not reasonable approximations. Not only are the elastic and viscous properties of polymer melts and solutions usually nonlinear, but they exhibit a *combination* of viscous and elastic response the relative magnitudes of which depend on the temperature and the time scale of the experiment. This *viscoelastic* response is dramatically illustrated by "silly putty" (a silicone polymer). When bounced (stressed rapidly), it is highly elastic, recovering most of the potential energy it had before being dropped. If stuck on the wall, however (stressed over a long time period), it will slowly flow down the wall, albeit with a high viscosity, and will show little tendency to recover any deformation.

We limit ourselves here to two-dimensional deformations. A detailed three-dimensional treatment of rheology is beyond the scope of this book. Several excellent treatises are available.[1-5]

15.2 BASIC DEFINITIONS

We begin our treatment of rheology with a discussion of *purely viscous flow*. For our purposes, this is defined as a deformation process in which all the

---

*This word is sometimes misspelled with a "t" in place of the "r." After reading some of the current literature, one isn't always sure that this is an error.

applied mechanical energy is nonrecoverably dissipated as heat in the material; that is, it is all converted to heat through *viscous energy dissipation*. Purely viscous flow is in most cases a good approximation for dilute polymer solutions and often for concentrated solutions and melts where the stress on the material is not changing too rapidly, that is, where it has a chance to approach an equilibrium flow situation.

The *viscosity* of a material expresses its resistance to flow under a mechanical stress. It is defined quantitatively in terms of two basic parameters, the *shear stress* τ and the *shear rate* (more correctly, the rate of shear straining) $\dot{\gamma}$. These quantities are defined in Fig. 15.1. Consider a point in a laminar flow field (Fig. 15.1). A coordinate system is established such that the x axis (sometimes designated the 1 coordinate direction) is in the direction of flow, and the y axis (the 2 direction) is perpendicular to surfaces of constant fluid velocity; that is, it is parallel to the *velocity gradient*. The z axis (3 or neutral direction) is mutually perpendicular to the others. A flow field in which the velocity and its gradient are everywhere perpendicular is known as a *viscometric flow* (see reference 2 for a more rigorous definition), and from a practical standpoint it is a type that can be simply treated analytically. Fortunately, many laminar-flow situations encountered are, or at least can reasonably be approximated by viscometric flows. Other examples of viscometric flows are given in Chapter 16.

A fluid surface at y moves with a velocity $u = dx/dt$ in the x direction, while the surface at $y + dy$ has a velocity $u + du$. The displacement gradient, dx/dy, is

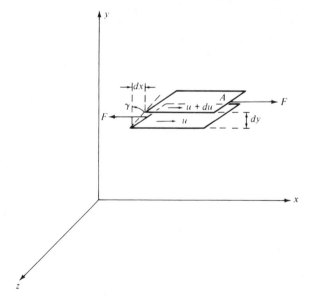

Figure 15.1   Definitions of shear stress and shear rate.

known as the *shear strain* and is given the symbol γ.

$$\gamma = \frac{dx}{dy} = \text{shear strain} \qquad \text{(dimensionless)} \qquad (15.1)$$

The *time rate of change* of shear strain $\dot{\gamma}$ (the dot is Newton's notation for the time derivative) is the so-called *shear rate*. Since the order in which the mixed second derivative is taken is immaterial (note that by sticking to two dimensions, we can write total rather than partial derivatives)

$$\dot{\gamma} = \frac{d}{dt}(\gamma) = \frac{d}{dt}\left(\frac{dx}{dy}\right) = \frac{d}{dy}\left(\frac{dx}{dt}\right) = \frac{du}{dy} \;(\text{time}^{-1}) \qquad (15.2)$$

Thus, an alternate definition of the shear rate is the velocity gradient du/dy.

The shear stress is simply the force (in the direction of flow) per unit area normal to the y axis:

$$\tau_{yx} = \frac{F \text{ (in } x \text{ direction)}}{A \text{ (normal to } y \text{ direction)}}\left(\frac{\text{force}}{\text{length}^2}\right) \qquad (15.3)$$

The subscript yx will henceforth be dropped unless specifically needed.

The *viscosity* η is *defined* as the ratio of shear stress to shear rate:

$$\eta \equiv \frac{\tau}{\dot{\gamma}} \qquad (15.4)$$

15.3   RELATIONS BETWEEN τ AND $\dot{\gamma}$ — FLOW CURVES

Newton's "law" of viscosity states that the shear stress is linearly proportional to the shear rate, the proportionality *constant* being the *viscosity* η.

$$\tau = \eta\dot{\gamma} \qquad (15.5)$$

(Most rheological work is done in the cgs system, with force in dynes, mass in grams, length in centimeters, and time in seconds. In this system, the unit of viscosity is dyne sec/cm^2 or the *poise*. All equations here are written with this system in mind. When using the English system of pound force, pound mass, feet, and seconds, each stress or pressure as written here must be multiplied by the dimensional *constant* $g_c = 32.2$ ft-lb$_m$/lb$_f$ sec^2). Fluids that obey this hypothesis are termed Newtonian. It holds quite well for low-molecular-weight fluids such as gases, water, and toluene. An arithmetic plot of τ versus $\dot{\gamma}$, a *flow curve*, is a straight line through the origin with a slope η for a Newtonian fluid (Fig. 15.2a). Taking logarithms of both sides of (15.5),

$$\log \tau = \log \eta + 1 \log \dot{\gamma}$$

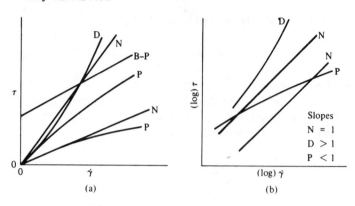

Figure 15.2 Types of flow curves: N, Newtonian; P, pseudoplastic; B-P, Bingham plastic (infinitely pseudoplastic); D, dilatant. (*a*) Arithmetic; (*b*) logarithmic.

Hence, a log–log plot of τ versus $\dot\gamma$, a logarithmic flow curve, will be a line of slope unity for a Newtonian fluid (Fig. 15.2*b*). This type of flow behavior would be expected for small molecules, where the structure, and therefore the resistance to flow, does not change with the intensity of shearing.

Unfortunately, many fluids do not obey Newton's hypothesis. Both *dilatant* (shear-thickening) and *pseudoplastic* (shear-thinning) fluids have been observed (Fig. 15.2). On logarithmic coordinates, dilatant flow curves have a slope greater than 1 and pseudoplastics a slope less than 1. Dilatant behavior is reported for certain slurries and implies an increased resistance to flow with intensified shearing. *Polymer melts and solutions are invariably pseudoplastic*; that is, their resistance to flow decreases with the intensity of shearing.

For non-Newtonian fluids, since τ is not directly proportional to $\dot\gamma$, the viscosity is not constant. Plots (or equations) giving η as a function of $\dot\gamma$ or τ are an entirely equivalent method of representing a material's equilibrium viscous shearing properties. A knowledge of the relation between any two of the three variables $\eta, \dot\gamma$, and τ completely defines the equilibrium viscous shearing behavior, since they are related by (15.4).

15.4   TIME-DEPENDENT BEHAVIOR

The types of non-Newtonian flow just described, though shear dependent, are time independent; that is, as long as a constant shear rate or stress is maintained, the same viscosity will be observed at equilibrium. Some fluids exhibit *reversible* time-dependent properties, however. When sheared at a constant rate or stress, the viscosity of a *thixotropic* fluid will decrease over a period of time (Fig. 15.3), implying a progressive breakdown of structure. If the shearing is stopped for a

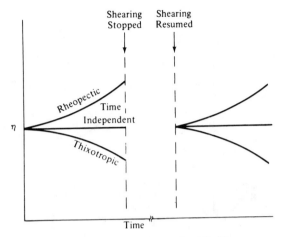

Figure 15.3 Time-dependent fluids.

while, the structure reforms, and the experiment may be duplicated. The ketchup that splashes all over after a period of vigorous tapping is a classic example. Thixotropic behavior is important in the paint industry, where smooth, even application with brush or roller is required, but it is desirable for the paint on the surface to "set up" to avoid drips and runs after application. The opposite sort of behavior is manifested by *rheopectic* fluids, for example, certain drilling muds used by the petroleum industry. When subjected to continuously increasing and then decreasing shearing, time-dependent fluids give flow curves as in Fig. 15.4.

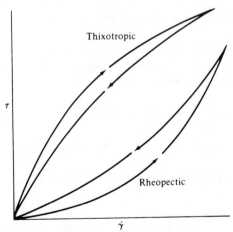

Figure 15.4 Flow curves for time-dependent fluids under continuously increasing and then decreasing shear.

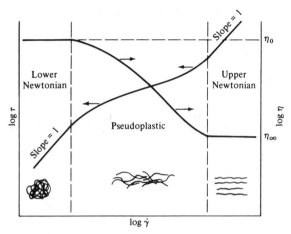

Figure 15.5   Generalized flow properties for polymer melts and solutions.

Once in a while, polymer systems will appear to be thixotropic or rheopectic. Careful checking (including before and after molecular-weight determination) invariably shows that the phenomenon is not reversible and that it is due to degradation or crosslinking of the polymer when in the viscometer for long periods of time, particularly at elevated temperatures. Other *transient* time-dependent effects in polymers are due to elasticity and are considered later, but for chemically stable polymer melts or solutions, the *equilibrium* viscous properties are time independent. We treat only such systems from here on.

15.5   POLYMER MELTS AND SOLUTIONS

When the flow properties of polymer melts and solutions can be measured over a wide enough range of shearing, the logarithmic flow curves appear as in Fig. 15.5. It is generally observed that

1.   At low shear rates (or stresses), a "lower Newtonian" region is reached with a so-called *zero-shear viscosity* η_0.
2.   Over several decades of intermediate shear rates, the material is pseudo-plastic.
3.   At very high shear rates, an "upper Newtonian" region with viscosity η_∞ is attained.

This behaviour can be rationalized in terms of molecular structure. At low shear, the randomizing effect of the thermal motion of the chain segments overcomes any tendency toward molecular alignment in the shear field. The molecules are thus in their most random and highly entangled state, and have their greatest resistance to slippage (flow). As the shear is increased, the molecules will begin to untangle and align in the shear field, reducing their resistance to slippage past one another. Under severe shearing they will be pretty much completely untangled and aligned, and reach a state of minimum resistance to flow. This is illustrated schematically in Fig. 15.5. Intense shearing will eventually lead to extensive breakage of main-chain bonds, that is, mechanical degradation.

It is worthwhile to consider here what happens to the highly oriented molecules when the shear field is removed. The randomizing effect of thermal energy tends to return them to their low-shear configurations, giving rise to an elastic retraction.

Some actual flow curves for polymer melts are shown in Fig. 15.6.[6] The data cover only a portion of the general range described above, as very few instruments are capable of obtaining data over the entire range.

15.6   QUANTITATIVE REPRESENTATION OF FLOW BEHAVIOR

In order to handle non-Newtonian flow analytically, it is desirable to have a mathematical expression relating τ and $\dot{\gamma}$, as Newton's "law" does for Newtonian fluids. A wide variety of such *constitutive relations* has been proposed, both theoretical and empirical.[1,7,8] All appear to fit at least some experimental data over a limited range of shear rates, but in general, the more "adjustable parameters" in the equation, the better fit it provides. (There is an old saying that with six constants you can draw an elephant, and with a seventh make his trunk wave.) The mathematical complexity of the equations increases greatly with the number of parameters, soon outstripping the data available to establish the parameters, making the equations impractical for engineering calculations.

The most common engineering model for purely viscous non-Newtonian flow is the so-called *power law*:

$$\tau = K(\dot{\gamma})^n \qquad (15.6a)$$

This is a two-parameter model, the adjustable parameters being K, the *consistency*, and n, the *flow index*. As written above, the dimensions of K depend on the magnitude of n (dimensionless); so strictly speaking, the power law should be (but rarely is) written

$$\tau = K|\dot{\gamma}|^{n-1}\dot{\gamma} \qquad (15.6b)$$

This way, K has the usual viscosity units.

On $\log \tau$ versus $\log \dot{\gamma}$ coordinates, a power-law fluid is represented by a straight line with slope n. Thus, for $n = 1$, it reduces to Newton's law, for $n < 1$,

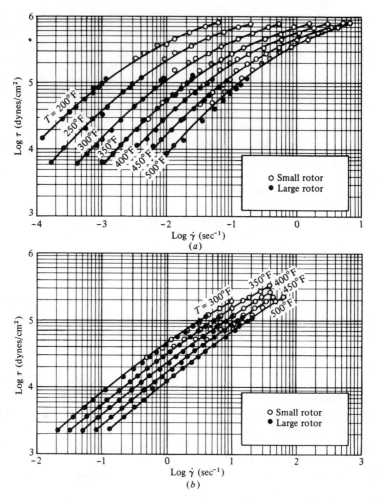

Figure 15.6 Flow curves for polymer melts;[6] (a) L-80 polyisobutylene; (b) low-density polyethylene.

the fluid is pseudoplastic; and for $n > 1$, dilatant. It can reasonably approximate only portions of actual flow curves over one or two decades of shear rate (see Fig. 15.6), but it does so with fair mathematical simplicity and is adequate for many engineering purposes. Many useful relations are obtained simply by replacing Newton's law with the power law in the usual fluid-dynamic equations.

Example 1. Determine the equation relating viscosity to shear rate for a power-law fluid.

Solution

$$\eta \equiv \frac{\tau}{\dot{\gamma}} = \frac{K\dot{\gamma}^n}{\dot{\gamma}} = K(\dot{\gamma})^{n-1} \qquad (15.7)$$

Thus a log–log plot of η versus $\dot{\gamma}$ for a power-law fluid will be linear with a slope of $n-1$. This points up an interesting limitation of the power law. Where the shear rate goes to zero (e.g., for Poiseuille flow at the center line of a cylindrical tube), the viscosity approaches infinity for a pseudoplastic power-law fluid, for which $n < 1$.

The Carreau model[1]

$$\frac{\eta - \eta_\infty}{\eta_0 - \eta_\infty} = [1 + (\lambda\dot{\gamma})^2]^{(n-1)/2} \qquad (15.8)$$

is a four-parameter model that is capable of representing all the features of the general flow curve in Fig. 15.5 and has been quite successful in fitting data for polymer melts and solutions over at least three or four decades of shear rate. The parameter λ is a time constant. In this day and age of the computer, four constants no longer seem excessive.

15.7   TEMPERATURE DEPENDENCE OF FLOW PROPERTIES

Also of engineering interest is the variation of flow properties with temperature. The *zero-shear* viscosity can often be represented by the relation

$$\eta_0 = Ae^{E/RT} \qquad (15.9)$$

over a range of several hundred degrees Fahrenheit above T_g, as can the viscosity of low-molecular-weight fluids. But polymer systems are rarely in the lower Newtonian range in commercial processing operations. Since the viscosity is a function of temperature and shear stress *or* shear rate,

$$\eta = f(\tau, T) \qquad or \qquad \eta = f'(\dot{\gamma}, T)$$

Over a similar temperature range, these functions are approximated by

$$\eta = Be^{E_\tau/RT} \qquad (\tau \text{ constant}) \qquad (15.10a)$$

$$\eta = Ce^{E_{\dot{\gamma}}/RT} \qquad (\dot{\gamma} \text{ constant}) \qquad (15.10b)$$

where E_τ is the activation energy for flow at constant shear stress and $E_{\dot{\gamma}}$ is the activation energy for flow at constant shear rate. Figure 15.7 shows plots of $\log \dot{\gamma}$ versus $1/T$ at constant τ.[6] Since $\dot{\gamma} = \tau/\eta$, the slope of these plots is $-E_\tau/R$.

Example 2.   What must the temperature be to halve the viscosity of the

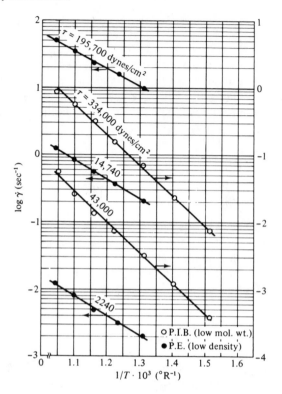

Figure 15.7   Temperature dependence of shear rate at constant shear stress.[6]

polymers in Fig. 15.7 at 300°F? The activation energies E_τ are 8830 cal/g mole for the polyisobutylene (PIB) and 5580 cal/g mole for the polyethylene (PE).

Solution.   At any value of shear stress τ, the viscosity is given by (15.10a). Therefore, between temperatures T_1 and T_2

$$\frac{\eta_2}{\eta_1} = \frac{e^{E_\tau/RT_2}}{e^{E_\tau/RT_1}} = \exp\left[\frac{E_\tau}{R}\left(\frac{1}{T_2} - \frac{1}{T_1}\right)\right]$$

or

$$\ln\frac{\eta_2}{\eta_1} = \frac{E_\tau}{R}\left(\frac{1}{T_2} - \frac{1}{T_1}\right)$$

In this case

$$\frac{\eta_2}{\eta_1} = \tfrac{1}{2}, \qquad \ln\tfrac{1}{2} = -0.693$$

For PIB

$$\frac{1}{T_2} - \frac{1}{T_1} = \frac{(-0.693)(1.99 \text{ cal/g mole } {}^\circ\text{K})}{8830 \text{ cal/g mole}} = -0.000156 \left(\frac{1}{{}^\circ\text{K}}\right)$$

$$-0.000156\left(\frac{1}{°K}\right) \times \left(\frac{1°K}{1.8°R}\right) = -0.000087 \left(\frac{1}{°R}\right)$$

$$\frac{1}{T_1} = \frac{1}{460 + 300} = 0.001316 \left(\frac{1}{°R}\right)$$

$$\frac{1}{T_2} = 0.001316 - 0.000087 = 0.001229 \left(\frac{1}{°R}\right)$$

$$T_2 = 813°R = 353°F$$

For PE

$$\frac{1}{T_2} - \frac{1}{T_1} = \frac{(-0.693)(1.99)}{5580} = 0.000247 \left(\frac{1}{°K}\right)$$

$$= -0.00137 \left(\frac{1}{°R}\right)$$

$$\frac{1}{T_2} = 0.001316 - 0.000137 = 0.001179\left(\frac{1}{°R}\right)$$

$$T_2 = 848°R = 388°F$$

The temperature is, therefore, an effective means of controlling melt viscosity in processing operations, but two drawbacks must be kept in mind: (a) it takes time and costs money to put in and take out thermal energy, and (b) excessive temperatures can lead to degradation of the polymer.

A more general expression for the variation of η_0 with temperature is obtained from the principle of corresponding states, which says that all amorphous polymers are mechanically equivalent equally removed from their glass transition temperatures. It is expressed quantitatively by the famous Williams–Landell–Ferry or WLF equation:

$$\log \frac{\eta_0(T)}{\eta_0(T_g)} = -\frac{17.44(T - T_g)}{51.6 + T - T_g} \tag{15.11}$$

Modifications of (15.11) that use different constants and characteristic temperatures other than T_g have been proposed.[9] But because T_g is extensively tabulated, (15.11) is of most practical value and does a pretty good job of representing experimental results for a wide variety of polymers in the temperature range of T_g to $T_g + 100°C$.

15.8   INFLUENCE OF MOLECULAR WEIGHT ON FLOW PROPERTIES

It has long been known that a polymer's molecular weight exerts a strong influence on the melt or concentrated solution viscosity. Experiments show that

$$\eta_0 \propto (\bar{M}_w)^1 \quad \text{for} \quad \bar{M}_w < \bar{M}_{w_c} \tag{15.12a}$$

$$\eta_0 \propto (\bar{M}_w)^{3.4} \quad \text{for} \quad \bar{M}_w > \bar{M}_{w_c} \tag{15.12b}$$

where $\bar{M}_{w_c}$ is a critical average molecular weight, thought to be the point at which molecular entanglements begin to dominate the rate of slippage of the molecules. $\bar{M}_{w_c}$ depends on the temperature and polymer type, but most commercial polymers are well above $\bar{M}_{w_c}$.

Equation (15.12) holds quantitatively for just about all polymer melts. The addition of a low-molecular-weight solvent, of course, cuts down entanglements and raises $\bar{M}_{w_c}$. Nevertheless, even moderately concentrated (say 25% or more) polymer solutions have viscosities proportional to $\bar{M}_w^{3.4}$ provided $\bar{M}_w$ is in the range of commercial importance. As the shear rate is increased, the number of

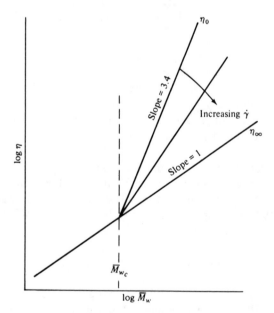

Figure 15.8   The effect of molecular weight on viscosity.

entanglements between chains is reduced, and, as expected, the dependence of viscosity on molecular weight decreases (Fig. 15.8).

Example 3.   By what percentage must $\bar{M}_w$ be changed to cut η_0 in half?

Solution.   From (15.12b)

$$\frac{\eta_{0_2}}{\eta_{0_1}} = \left(\frac{\bar{M}_{w_2}}{\bar{M}_{w_1}}\right)^{3.4}$$

$$\frac{\bar{M}_{w_2}}{\bar{M}_{w_1}} = \left(\frac{\eta_{0_2}}{\eta_{0_1}}\right)^{1/3.4} = (\tfrac{1}{2})^{1/3.4} = 0.816$$

$$\% \text{ change} = 100\left(\frac{\bar{M}_{w_2} - \bar{M}_{w_1}}{\bar{M}_{w_1}}\right) = 100\left(\frac{\bar{M}_{w_2}}{\bar{M}_{w_1}} - 1\right) = -18.4\%$$

Thus it only takes an 18% decrease in chain length to halve the melt viscosity. Although things won't be quite so dramatic at high shear rates, this illustrates the importance of controlling molecular weight to achieve the desired processing properties.

15.9   THE EFFECTS OF PRESSURE ON VISCOSITY

For all fluids, viscosity increases with increasing pressure as the free volume, and hence the ease of molecular slippage, is decreased. With liquids, including polymer melts, because of their relative incompressibility, the effect becomes noticeable only at very high pressures (many thousand pounds per square inch). The available data are fitted by an equation of the type

$$\eta = Ae^{BP} \tag{15.13}$$

where A and B are constants.

Carley[10] has critically reviewed work on polymer melts and concludes that pressure effects are of minor significance in most processing situations, provided the temperature is not too close to a transition. High pressures raise both T_g and T_m slightly, and of course the viscosity shoots up tremendously as either is reached.

REFERENCES

1. R. B. Bird, R. C. Armstrong, and O. A. Hassager, *Dynamics of Polymeric Liquids*, Vol. I: Fluid Mechanics, Wiley, New York, 1977.

2. B. D. Coleman, H. Markovitz, and W. Noll, *Viscometric Flows of Non-Newtonian Fluids*, Springer-Verlag, New York, 1966.

3. R. M. Christensen, *Theory of Viscoelasticity. An Introduction*, Academic, New York, 1971.

4. R. Darby, *Viscoelastic Fluids*, Marcel Dekker, New York, 1976.

5. S. Middleman, *The Flow of High Polymers*, Interscience, New York, 1968.

6. D. M. Best and S. L. Rosen, *Polym. Eng. Sci.*, 8, 1, 116 (1968).

7. R. B. Bird, W. E. Stewart, and E. N. Lightfoot, *Transport Phenomena*, Wiley, New York, 1960, Chapter 1.

8. F. Rodriguez, *Principles of Polymer Systems*, McGraw-Hill, New York, 1970, Chapter 7.

9. A. V. Tobolsky, *Properties and Structure of Polymers*, Wiley, New York, 1960, Chapter 8.

10. J. F. Carley, *Mod. Plast.*, 39, 4, 123 (1961).

CHAPTER 16

Viscometry and Tube Flow

16.1 INTRODUCTION

This chapter considers some of the techniques used to establish the viscous flow properties discussed in the previous chapter. One of the common methods, Poiseuille (laminar) flow in cylindrical tubes, is also important from a technological standpoint, as polymer melts and solutions are often transported and processed in this fashion. Because of their tremendous viscosities, laminar flow is far more prevalent with polymeric fluids than with nonpolymer fluids; in fact, turbulence is normally encountered only with dilute (less than a few weight percent, or so) polymer solutions.

16.2 VISCOUS ENERGY DISSIPATION

Regardless of the viscometric technique, determination of the true isothermal flow properties at high shear rates can be very difficult because of the high rates of viscous energy dissipation, which makes it hard to maintain isothermal conditions. The *rate of viscous energy dissipation per unit volume* $\dot{E}$ is

$$\dot{E} = \tau\left(\frac{\text{dynes}}{\text{cm}^2}\right)\dot{\gamma}\left(\frac{1}{\text{sec}}\right) \times 1\left(\frac{\text{cm}}{\text{cm}}\right) = \tau\dot{\gamma}\left(\frac{\text{dyne cm}}{\text{cm}^3\ \text{sec}}\right) = \tau\dot{\gamma}\left(\frac{\text{ergs}}{\text{cm}^3\ \text{sec}}\right) \qquad (16.1)$$

Example 1. Obtain an expression for the adiabatic rate of temperature rise in a polymer sample subjected to a shear stress τ and shear rate γ.

Solution

$$\frac{dT}{dt} = \tau\dot{\gamma}\left(\frac{\text{ergs}}{\text{cm}^3 \text{ sec}}\right) \times \frac{1}{\rho\,(\text{g/cm}^3)\,C_p\,(\text{ergs/g}\,^\circ\text{C})} = \frac{\tau\dot{\gamma}}{\rho C_p}\left(\frac{^\circ\text{C}}{\text{sec}}\right) \quad (16.2)$$

Not only is viscous energy dissipation an important consideration in viscometry, where great care must be taken in the design of viscometers to permit adequate temperature regulation, but it also must be taken into account in the design of processing systems. In the steady-state operation of extruders, for example, virtually all the energy required to melt and maintain the polymer in the molten state is supplied by the mechanical drive. Here, however, we limit our considerations to isothermal flow situations.

16.3   POISEUILLE FLOW

Axial laminar (Poiseuille) flow in a tube of cylindrical cross section is an example of the viscometric flow field discussed in the previous chapter. Here, the geometry of the situation dictates the use of cylindrical coordinates. Fluid motion is in the x (or 1) direction along the tube axis, the velocity gradient is everywhere directed in the outward radial r (or 2) direction, and the mutually perpendicular or neutral direction is the θ (or 3) coordinate. If the fluid pressure is a function of distance along the tube (x coordinate) only, equating the shear force on the surface of a cylindrical fluid element of radius r and length dx to the axial forces on it arising from the differential pressure drop $-dP$ across its ends (Fig. 16.1) (they must balance in an equilibrium flow situation),

$$\underset{\text{surface shear force}}{2\pi r\, dx\, \tau_{rx}} = \underset{\text{net pressure force}}{-\pi r^2\, dP} \quad (16.3)$$

Thus the dependence of shear stress on radius (denoted $\tau(r)$) is

$$\tau_{rx}(r) = -\frac{r}{2}\left(\frac{dP}{dx}\right) \quad (16.4)$$

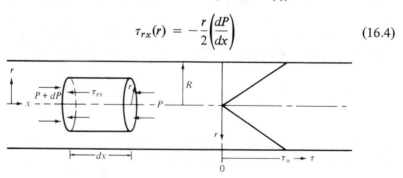

Figure 16.1   Force balance on an element in a cylindrical tube.

For a tube with inner radius R (dropping the subscripts on shear stress)

$$\tau_w = -\frac{R}{2}\left(\frac{dP}{dx}\right) \tag{16.5}$$

where τ_w is the shear stress at the tube wall, and

$$\tau(r) = \left(\frac{r}{R}\right)\tau_w \tag{16.6}$$

So the shear stress varies linearly with radius from zero at the tube center to a maximum of $\tau_w = -(R/2)(dP/dx)$ at the tube wall. *Note that this result does not depend in any way on the fluid properties.*

Since the axial fluid velocity u is a function of radial position only (denoted $u(r)$), we may write

$$du(r) = \left(\frac{du}{dr}\right)dr \tag{16.7}$$

Using the boundary conditions $u(R) = 0$ (i.e., the fluid sticks to the wall) and u at radius $r = u(r)$, and integrating,

$$u(r) = \int_0^{u(r)} du(r) = \int_R^r \left(\frac{du}{dr}\right)dr \tag{16.8}$$

Realizing that the velocity gradient (du/dr) is simply the shear rate $\dot{\gamma}$,

$$u(r) = \int_R^r \dot{\gamma}(\tau)dr \tag{16.9}$$

where τ, in turn, is a function of r (see (16.4)). The function $\dot{\gamma}(\tau)$ is, of course, the flow curve for the material.

So, for a given pressure gradient, the velocity profile in the tube may always be calculated from the flow curve by

1. Choosing an r and calculating τ from (16.4),
2. Obtaining $\dot{\gamma}$ at the τ above from the flow curve,
3. Plotting $\dot{\gamma}$ versus r (or otherwise numerically representing the relation),
4. Integrating from R to r, which gives u at r.

If an analytic representation of the flow curve is available, (16.9) may be integrated directly (provided the representation is simple enough). For example (verification will be left as an exercise for the reader, as they say in the math

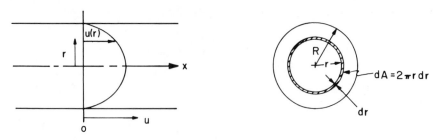

Figure 16.2   Determination of volumetric flow rate.

books), for a power-law fluid $\dot{\gamma}(\tau) = (\tau/K)^{1/n}$, and

$$u(r) = \frac{[(-dP/dx)/2K]^{1/n}}{(1/n) + 1} [R^{(1/n)+1} - r^{(1/n)+1}] \tag{16.10}$$

Equation (16.10) reduces to the usual Newtonian parabolic profile for $n = 1$.

Looking at a differential ring of the tube cross section, with thickness dr, located at radius r where the velocity is $u(r)$ (Fig. 16.2), the differential volumetric flow rate is

$$dQ = \underbrace{u(r)}_{\text{local velocity}} \quad \underbrace{2\pi r \, dr}_{\text{area of differential ring}} \tag{16.11}$$

Integrating over the tube cross section,

$$Q = 2\pi \int_0^R u(r) r \, dr \left(\frac{\text{cm}^3}{\text{sec}}\right) \tag{16.12}$$

So, knowing the velocity profile allows calculation of the volumetric throughput Q. For a power-law fluid,

$$Q = \left[\frac{-(dP/dx)}{2K}\right]^{1/n} \left(\frac{\pi}{(1/n) + 3}\right) R^{(1/n)+3} \tag{16.13}$$

The *average velocity* in the tube is defined by

$$V \equiv \frac{Q}{\pi R^2} \tag{16.14}$$

Velocity profiles for power-law fluids in a tube are plotted in dimensionless form in Fig. 16.3 for several values of n. The greater the degree of pseudoplasticity (i.e., the lower n), the flatter the profile becomes. For the Newtonian fluid, $n = 1$, the profile is parabolic.

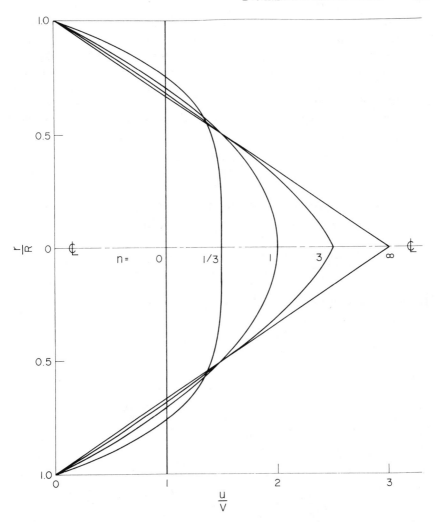

Figure 16.3   Velocity profiles for the laminar flow of power-law fluids in tubes.

16.4   DETERMINATION OF FLOW CURVES

The formulas just developed allow the relation of pressure gradient, velocity profile, and volumetric flow rate *as long as the shear stress–shear rate relation* (flow curve) *for the fluid is known.* The problem in viscometry is just the reverse; that is, how is the flow curve obtained from pressure drop–volumetric flow rate measurements in a cylindrical tube? *If* the mathematical form of the flow curve, that is, a particular constitutive equation *is assumed a priori*, the integrated equations as developed above may be used to establish the parameters in the constitutive relation. For example, if it is assumed that the power law represents the flow curve of a fluid under investigation, two readings of dP/dx versus Q in a tube of known R will allow calculation of K and n from (16.13). However, in the general case, the form of the constitutive equation is not known a priori and must be established by viscometry. This may be done first by integrating (16.12) by parts:

$$Q = 2\pi \left[\frac{r^2 u(r)}{2} \Big|_0^R - \int_0^R \frac{r^2}{2} du(r) \right] \tag{16.15}$$

Fortunately, the first term in the brackets of (16.15) is zero because at the upper limit, $u(R) = 0$, and at the lower limit, $r = 0$. Therefore,

$$Q = -\pi \int_0^R r^2 \, du(r) = -\pi \int_0^R r^2 \left(\frac{du(r)}{dr} \right) dr = -\pi \int_0^R r^2 \dot{\gamma}(r) dr \tag{16.16}$$

Using (16.6) to change the independent variable from r to τ, recalling that at $r = 0$, $\tau = 0$, and at $r = R$, $\tau = \tau_w$,

$$\frac{\tau_w^3 Q}{\pi R^3} = -\int_0^{\tau_w} \dot{\gamma}(\tau) \tau^2 \, d\tau \tag{16.17}$$

Applying Leibnitz's rule to differentiate both sides of (16.17) with respect to τ_w gives

$$\frac{1}{\pi R^3} \left[\tau_w^3 \frac{dQ}{d\tau_w} + 3\tau_w^2 Q \right] = -\dot{\gamma}_w \tau_w^2 \tag{16.18}$$

Or

$$-\dot{\gamma}_w = \frac{1}{\pi R^3} \left[\tau_w \frac{dQ}{d\tau_w} + 3Q \right] \tag{16.19}$$

Since $d \ln \xi = d\xi/\xi$ (ξ = any variable),

$$-\dot{\gamma}_w = \frac{1}{\pi R^3} \left[3Q + \frac{Qd \ln dQ}{d \ln \tau_w} \right] \tag{16.20}$$

Or

$$-\dot{\gamma}_w = \frac{4Q}{\pi R^3} \left[\frac{3}{4} + \frac{1}{4} \frac{d \ln Q}{d \ln \tau_w} \right] \tag{16.21}$$

$$-\dot{\gamma}_w = \Gamma \left[\frac{3}{4} + \frac{1}{4} \frac{d \ln \Gamma}{d \ln \tau_w} \right] \tag{16.22}$$

where

$$\Gamma \equiv \frac{4Q}{\pi R^3} = \frac{8V}{D} \tag{16.23}$$

is known as the *apparent* shear rate.

Equations (16.21) and (16.22) are equivalent forms of the *Rabinowitsch equation*, allowing calculation of the shear rate at the tube wall from experimental data. Note the necessity of obtaining the slope of a log Γ versus log τ_w plot (or otherwise differentiating the data) to obtain $\dot\gamma_w$. For a single tube, Γ is proportional to Q and τ_w to dP/dx; so $d \ln Q/d \ln (dP/dx)$ may be substituted for $d \ln \Gamma/d \ln \tau_w$ in (16.22). However, (16.22) allows data from several different tubes, if available, to be combined, providing more accurate values for the slopes. Values of $\dot\gamma_w$ are then combined with values of τ_w calculated from (16.5) to give the flow curve.

Neglect of the Rabinowitsch correction (the term in brackets in (16.21) and (16.22)) can lead to serious error. To illustrate, differentiation of (16.13) and substitution of (16.5) and (16.23) reveal that for a power-law fluid, $d \ln \Gamma/d \ln \tau_w = 1/n$, a constant. The Rabinowitsch equation then becomes

$$-\dot\gamma_w = \Gamma \left[\frac{3n + 1}{4n} \right] \qquad \text{(power-law fluids)} \tag{16.24}$$

Only for a Newtonian fluid, $n = 1$, is Γ equal to the *true* shear rate at the tube wall $\dot\gamma_w$. For an n of $1/3$, a typical value for a polymer melt or solution, the correction term is 1.5; so the apparent shear rate is 50% lower than the true value at the tube wall. The Rabinowitsch correction accounts for the fact that the velocity gradient (shear rate) is greater at the tube wall for pseudoplastic fluids than for a Newtonian fluid at a given Q, as illustrated in Fig. 16.3.

When comparing viscometric data from tubes with those obtained from other types of viscometers, and when using tube-flow data to predict behavior in other geometries, the use of the Rabinowitsch correction is essential to give the true shear stress–shear rate relation. If all we are concerned with is flow in cylindrical tubes, however, it is not necessary. A look at (16.17) shows that for the flow of a given fluid in cylindrical tubes, τ_w is a unique function of the apparent shear rate Γ, because the value of the integral depends only on τ_w, regardless of the nature of the $\dot\gamma(\tau)$ relation (flow curve) of the fluid. Thus experimental determinations of τ_w versus Γ obtained from cylindrical tubes are applicable for scaleup purposes to other cylindrical tubes through which the same fluid is flowing (see reference 1, Example 3.2).

The preceding analysis of flow in tubes has assumed that the flow properties of the fluid are independent of pressure. This is probably a reasonable assumption in most cases. However, as mentioned in Section 15.7, viscosity increases with pressure, particularly as T_g or T_m is approached, and if the pressure drop across the tube is large, corrections may be necessary. Such calculations have been discussed.[2]

16.5 ENTRANCE CORRECTIONS

A capillary viscometer is a device in which the fluid under investigation is forced from a reservoir through a cylindrical capillary tube. Viscometers are operated with either flow rate or pressure as the independent variable. In the former case, the fluid is usually driven by a piston advancing through the reservoir at a known constant rate with the force on the piston recorded; in the latter case, regulated gas pressure drives the fluid and the volumetric flow rate is measured. Unfortunately, calculating the equilibrium pressure gradient from the data so obtained is not always a simple matter of setting $dP/dx = \Delta P/L$, where ΔP is the measured overall pressure drop across the capillary of length L. The gradient does not become constant until equilibrium flow is reached, some distance downstream from the entrance of the capillary, as illustrated in Fig. 16.4, and approximating the true equilibrium gradient dP/dx with the measured $\Delta P/L$ can cause considerable error. Higher-than-equilibrium pressure losses are observed in the entry region because (a) kinetic energy must be added to the fluid as it is accelerated from low velocities in the reservoir to higher velocities in the capillary, (b) rearrangement of the velocity profile dissipates additional energy, and (c) a viscoelastic fluid *stores* some energy elastically when going from the low stress reservoir to the high stress capillary (think of stretching a rubber band). Recovery of this stored energy at the tube exit leads to such things as die swell, melt fracture, and other "anomalies" observed with the flow of viscoelastic fluids.

Bagley[3] has developed a method for getting the true pressure gradient from

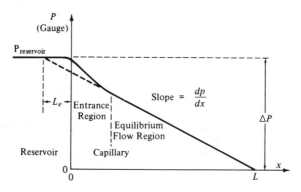

Figure 16.4   Pressure profile in a capillary viscometer.

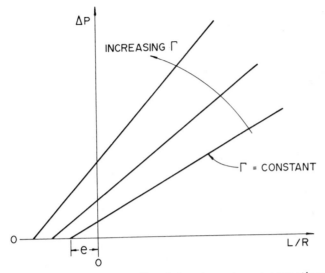

Figure 16.5   "Bagley plot" to determine entrance corrections.

capillary viscometer data. As seen in Fig. 16.4,

$$\left(\frac{dP}{dx}\right) = \frac{\Delta P}{(L + L_e)} = \frac{\Delta P}{(L + eR)} \qquad (16.25)$$

where L_e is a fictitious "entrance length." L_e is commonly expressed as the product of e, the *entrance correction*, and the tube radius R. Rewriting (16.5), dividing numerator and denominator by R, and recalling that τ_w is a function of Γ *only*,

$$\tau_w = \frac{R\Delta P}{2(L + eR)} = \frac{\Delta P}{2[(L/R) + e]} = f(\Gamma) \qquad (16.26)$$

where f denotes "function of." Rearranging,

$$\Delta P = 2\left(\frac{L}{R} + e\right)f(\Gamma) \qquad (16.27)$$

Thus, if a series of experiments is run in which the capillary L/R ratio is varied, a *Bagley plot* of ΔP versus L/R at constant $\Gamma = 4Q/\pi R^3$ should give a series of straight lines, one for each constant value of Γ. The $\Delta P = 0$ intercept is the entrance correction $-e$, (Fig. 16.5). The various intercepts at constant apparent shear rates give e as a function of τ_w or $\dot{\gamma}_w$. The need to use several capillaries greatly increases the work, but for melts, it appears that $(L/R) \gg e$ (i.e., the entrance correction becomes negligible) only for $(L/R) > 100$ or so. For solutions at high flow rates the correction may be significant at much higher L/R's.

The reverse of this procedure may be important in design situations; that is, the entrance correction must be known as a function τ_w and included in the overall pressure drop calculations. In extruder dies and spinerettes for fiber spinning, the L/R's rarely exceed 10, and the entrance loss may be the major contribution to the overall pressure drop. Carley[4] provides examples of design calculations involving these concepts.

Once the entrance correction has been established, it can be used to calculate the true equilibrium gradient through (16.25), which in turn is used in (16.5) and (16.21) or (16.22) to obtain the flow curve, relating the shear stress to shear rate, both at the tube wall. In a few instances other minor corrections have been applied, for example, considering the pressure drop in the viscometer reservoir.[5]

16.6   THE COUETTE VISCOMETER

Another common device for measuring viscous properties is the cup-and-bob or Couette viscometer, a diagram of which is given in Fig. 16.6. The fluid is confined in the gap between two concentric cylinders, one of which moves relative to the other at a known angular velocity while the torque on one is measured. This is another classic example of a viscometric flow, the 1 coordinate being the tangential or θ direction and the 2 coordinate the radial direction.

Example 2

a.   Neglecting end effects, determine the shear stress as function of radius in terms of the measured torque M on the stationary inner cylinder (bob) and

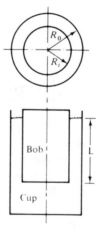

Figure 16.6   Schematic of Couette (cup-and-bob) viscometer.

of the geometry of the apparatus, as the outer cylinder (cup) is rotated with an angular velocity ω (radians per second).

b.   The problem of determining the shear rate as a function of radius is not an easy one. Bird, Stewart, and Lightfoot[6] do it by *assuming* a Newtonian relation between shear stress and shear rate. In general, however, the nature of the flow curve is not known a priori and must be determined by viscometry. One means of doing this is to make the gap between the cylinder $(R_0 - R_i)$ very small compared to the radius of either cylinder. Letting $(R_0 - R_i) = d$ and $R_0 \cong R_i = R$, obtain the expression for shear rate in terms of ω and the geometry.

Solution

a.   In a rotating system at equilibrium, Σ torques $= 0$, or there would be angular acceleration. Consider a ring of fluid with inner radius R_i and outer radius r: $M(r) = M(R_i)$

$$\underbrace{\underbrace{(2\pi r L)\,(\tau(r))}_{\substack{\text{surface} \\ \text{area}}}\ \underbrace{(r)}_{\substack{\text{moment} \\ \text{arm}}}}_{\text{force}} = M(R_i)$$

$$\tau(r) = \frac{M}{2\pi r^2 L} \tag{16.28}$$

b.   This situation approximates the case of two flat plates, separated by a distance d, one sliding past the other with a linear velocity equal to the tangential velocity $R\omega$

$$\dot{\gamma} = \frac{R\omega}{d} \qquad \text{(for } d \ll R_i \text{ only)} \tag{16.29}$$

Where the geometric approximations in Example 2b above are not applicable, Kreiger and Maron[7] have presented an analysis similar to the Rabinowitsch development for flow in tubes. It involves differentiation of the M versus ω data but, unfortunately, is in the form of an infinite series. If $R_0/R_i < 1.2$, a closed-form approximation is available, however.

Obviously, the analysis above is not valid in the area beneath the bob at the bottom of the viscometer. This is best taken into account by making measurements with two fluid depths, the lower being well above the bottom of the bob, and using the *differences* between the torques and depths in (16.28), thereby subtracting out the effects of non-Couette flow. Another approach is illustrated in Example 4.

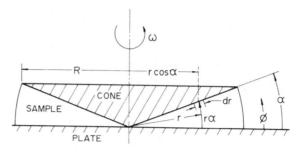

Figure 16.7   Schematic of cone-and-plate viscometer.

16.7   CONE-AND-PLATE VISCOMETER

Still another popular type of viscometer is the cone-and-plate, in which the sample is sheared between a flat plate and a broad cone whose apex contacts the plate (Fig. 16.7). For small cone-plate angles α, this approximates a viscometric flow with (in spherical coordinates) flow in the tangential θ or 1 direction and the gradient in the azimuthal ϕ or 2 direction. Here, the radial direction is the neutral or 3 coordinate.* It may be shown that true viscometric flow of this type is inconsistent with the equations of motion if the inertial terms are included. (Formal solutions to flow problems in rotational viscometers normally neglect the inertial terms in the Navier-Stokes equations, as we tacitly do here.) There must, therefore, be radial and azimuthal velocity components. These are minimized in practice by keeping α quite small — often less than 1°. The great advantage of this type of device is that (for small α) the shear rate, and hence the shear stress, is constant throughout the gap.

Example 3.   Obtain the expressions for shear rate and shear stress in a cone-and-plate viscometer in terms of the rate of cone rotation ω, the measured torque M, and the geometry.

Solution (See Fig. 16.7).   The tangential velocity v of a point on the cone relative to the plate is $v = \omega r \cos \alpha$. Fluid is sheared between that point and the plate over a distance $d = \alpha r$:

$$\dot{\gamma} = \frac{v}{d} = \frac{\omega r \cos \alpha}{\alpha r} \xrightarrow[\alpha]{\text{small}} \frac{\omega}{\alpha} \qquad (\textit{independent of } r) \qquad (16.30)$$

---

*At first glance, it might seem that there is a gradient component in the r direction because tangential velocity increases with r. However, material points in a cone at constant ϕ do not move *relative* to one another; that is, they undergo rigid-body rotation, and so there is no shearing in the r direction.

$$\text{torque} = \underbrace{(\text{shear stress})(\text{area})}_{\text{force}}(\text{moment arm})$$

$$dM = (\tau)(2\pi r\, dr)(r \cos \alpha)$$

Integrating over the cone face

$$M = 2\pi\tau \cos \alpha \int_{0}^{R/\cos \alpha} r^2\, dr = \frac{2\pi\tau R^3}{3 \cos^2\alpha} \xrightarrow[\alpha]{\text{small}} \frac{2\pi\tau R^3}{3}$$

$$\tau = \frac{3M}{2\pi R^3} \tag{16.31}$$

Note: Here, τ is constant and can be taken outside the integral, because $\tau = f(\dot{\gamma})$ and $\dot{\gamma}$ is constant, as shown by (16.30).

Cone-and-plate viscometers of the type shown here are usually limited to fairly low shear rates. At higher shear rates solutions tend to be flung from the gap by centrifugal force, and melts tend to "ball up" (like rubbing a finger over dry rubber cement). These problems can be overcome by enclosing the fluid around a biconical rotor, giving, in effect, two cone-and-plate viscometers back to back.[8] The flow curves in Chapter 15 were obtained with such a device.

Example 4.   One means of minimizing end effects in a Couette viscometer is to make the bottom of the bob a cone, the apex of which contacts the cup, so that the area beneath the bob is a cone-and-plate viscometer. For a Couette geometry in which the gap d is much smaller than the bob radius R_i, what must the cone angle α be to match the shear rates in the Couette and cone-and-plate regions?

Solution.   By equating the expressions for shear rate from Examples 2b and 3, $\alpha = d/R$.

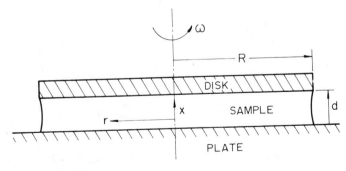

Figure 16.8   Schematic of disk-plate viscometer.

16.8   DISK–PLATE VISCOMETER

Another type of viscometer which finds occasional use is the disk–plate vis-cometer (Fig. 16.8). A disk of radius R rotates with an angular velocity of ω relative to a parallel fixed plate. The disk and plate are separated by a distance d $(d \ll R)$, with the test fluid in between. The torque M on either the disk or plate is measured. This is known as torsional flow and is another example of a viscometric flow.

Example 5

a.   Identify the 1 (flow), 2 (gradient), and 3 (neutral) directions with respect to the cylindrical coordinates in Fig. 16.8.*

b.   Obtain an expression for $\dot{\gamma}$. Why is the cone-and-plate geometry usually preferred?

c.   When a power-law fluid is in the gap, obtain an expression for the torque M needed to maintain a steady rate of rotation ω.

Solution

a.   1 (flow) $= \theta$, 2 (gradient) $= x$, and 3 (neutral) $= r$.

b.
$$\dot{\gamma} = \frac{\text{relative velocity, disk to plate}}{\text{separation distance}} = \frac{r\omega}{d} \qquad (16.32)$$

In this geometry, unlike the cone and plate, the shear rate varies linearly with radius, and so is not uniform throughout the fluid.

c.   Consider a differential ring of the disk (or plate) surface, at radius r with thickness dr:

$$dA = 2\pi r\, dr$$

$$dM = r\tau\, dA = 2\pi r^2 \tau\, dr$$

For a power-law fluid, $\tau = K\dot{\gamma}^n = K(r\omega/d)^n$,

$$M = 2\pi K \left(\frac{\omega}{d}\right)^n \int_0^R r^{2+n}\, dr = \frac{2\pi K(\omega/d)^n}{3+n} R^{3+n}$$

or,

$$M = \frac{2\pi K R^3}{3+n}(\dot{\gamma}_R)^n \qquad \text{(power-law fluid)} \qquad (16.33)$$

where

$$\dot{\gamma}_R \equiv \frac{R\omega}{d} \qquad (16.34)$$

When the form of the flow curve is not known a priori, it may be obtained

*See footnote to Section 16.7. Here, points at constant x undergo rigid-body rotation.

through a Rabinowitsch-like development that relates shear stress and shear rate at $r = R$:

$$\tau_R = \frac{3M}{2\pi R^3}\left(1 + \frac{1}{3}\frac{d\ln M}{d\ln \dot\gamma_R}\right) \tag{16.35}$$

16.9   TURBULENT FLOW

On occasions, turbulent flow is encountered with dilute polymer solutions. As with the turbulent flow of Newtonian fluids, pressure drops are conveniently handled in terms of the Fanning friction factor

$$f = \frac{Rg_c}{\rho V^2}\left(\frac{dP}{dx}\right) \tag{16.36}$$

(The dimensional constant $g_c = 32.2\,\text{ft-lb}_m/\text{lb}_f\sec^2$ is included here because these equations are often used with the English engineering system of units.) Since $\tau_w = (R/2)(dP/dx)$ (the minus sign has been dropped, it being understood that flow is in the direction of decreasing pressure),

$$f = \frac{2\tau_w g_c}{\rho V^2} \tag{16.37}$$

Equations (16.36) and (16.37) are in no way dependent on the nature of the fluid. If, as suggested by Metzner and Reed,[9] the analogy between the *laminar flow* of non-Newtonian and Newtonian fluids is to be preserved, then the usual relation

$$f = \frac{16}{\text{Re}} \tag{16.38}$$

where Re is the Reynolds number, is applicable to both. Combining (16.37) and (16.38), the *generalized Reynolds number* — for any fluid — becomes

$$\text{Re} = \frac{8\rho V^2}{\tau_w g_c} \tag{16.39}$$

For a power-law fluid, $\tau g_c = K(\dot\gamma)^n$, (16.5), (16.13), and (16.14) give

$$\text{Re} = \frac{8D^n V^{2-n}\rho}{K[6 + (2/n)]^n} \tag{16.40}$$

Actually, Metzner and Reed use an alternate formulation. Since the shear stress at the tube wall is a unique function of the apparent shear rate for laminar flow in tubes, the power law may be written at the tube wall with the aid of (16.24) as

$$\tau_w g_c = K(\dot{\gamma}_w)^n = K\left(\Gamma\left[\frac{3n+1}{4n}\right]\right)^n = K'\Gamma^n \tag{16.41}$$

where

$$K' = K\left[\frac{3n+1}{4n}\right]^n \tag{16.42}$$

With these variables, the generalized Reynolds number becomes

$$Re = \frac{D^n V^{2-n}\rho}{K'8^{n-1}} \tag{16.43}$$

Both (16.40) and (16.43) reduce to the usual Newtonian relation, $Re = DV\rho/\eta$, with $\eta = K = K'$, when $n = 1$.

Equations (16.36)–(16.39) *by definition* should describe the behavior of all fluids in *equilibrium* laminar viscous flow. They do, *provided that the true equilibrium pressure gradient* is used in (16.36). Approximation of the equilibrium gradient by $\Delta P/L$ can cause considerable error.

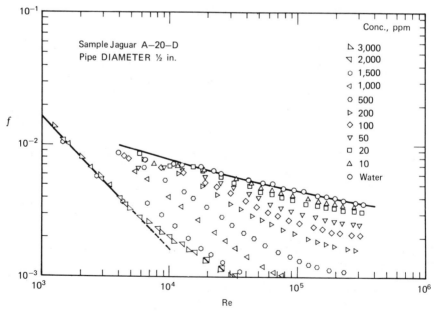

Figure 16.9   Turbulent drag reduction.[13] Copyright 1972 by the AMERICAN CHEMICAL SOCIETY. Reprinted with permission of the copyright owner.

16.10 DRAG REDUCTION[10-12]

After having defined the Reynolds number for non-Newtonian fluids, it would be nice to be able to say that one could then use the standard friction factor—Reynolds number chart to make the usual turbulent-flow calculations. Such is not always the case, however, because of a startling phenomenon known as *drag reduction*, which is often observed in the turbulent flow of non-Newtonian fluids (Fig. 16.9).[13] Small quantities of a polymeric solute can cut the friction factor significantly at higher Reynolds numbers, despite the fact that the solution viscosity is slightly greater than that of the solvent. For example, 200 ppm of guar gum. (a natural, water-soluble polymer) reduces the friction factor in Fig. 16.9 by approximately a factor of 5 at Re $= 10^5$, while the viscosity of the solution is 24% greater than that of pure water.

This phenomenon has important practical applications. Pressure drops, and therefore pumping costs, can be reduced for a given flow rate, or the capacity of a pumping system can be increased by the addition of a drag-reducing solute. Fire departments add poly(ethylene oxide) to their pumpers to increase capacity. Experiments have been conducted with the aim of increasing ship speed by squirting out a drag-reducing additive at the bow.

The causes of drag reduction are not yet clear. It has been suggested that it may be due to thickening of the laminar sublayer. Another possible cause is the viscoelasticity of polymer solutions (polymer solutions that seem relatively inelastic are more likely to be amenable to the usual treatment). Most of the energy dissipation in turbulent flow occurs in small, high-frequency eddies. At these frequencies (the reciprocals of which are smaller than the material's relaxation time, giving high Deborah numbers; see Chapter 18) the material responds elastically (like bouncing the silly putty instead of letting it flow down the wall), passing the stored elastic energy from eddy to eddy instead of dissipating it. Drag reduction is known to be most pronounced for solutions of high-molecular-weight, flexible-chain polymers, possibly because these form highly elastic solutions. Its magnitude is reduced markedly as the polymers are degraded by continued flow. As with Newtonians, the friction factor—Reynolds Number relation is a function of pipe roughness, but unlike the Newtonian case, the f—Re curve depends on pipe diameter. Regardless, the sad fact remains that at present there are no quantitative methods for predicting which polymeric solutes will produce drag reduction in a particular solvent, or over what range of concentrations it will be observed and what its magnitude will be.

REFERENCES

1. J. M. McKelvey, *Polymer Processing,* Wiley, New York, 1962, Chapter 3.
2. R. C. Pennwell, R. S. Porter, and S. Middleman, *J. Polym. Sci.* A-2, 9, 731 (1971).

3. E. B. Bagley, *J. Appl. Phys.*, **28**, 624 (1957).

4. J. F. Carley, *Soc. Plast. Eng. J.*, **19**, 12, 1263 (1963).

5. Van Wazer et al., *Viscosity and Flow Measurement – A Laboratory Handbook of Rheology*, Interscience, New York, 1963.

6. R. B. Bird, W. E. Stewart, and E. N. Lightfoot, *Transport Phenomena*, Wiley, New York, 1960, p. 94, Example 3.5-1.

7. I. M. Kreiger and S. H. Maron, *J. Appl. Phys.*, **25**, 72 (1954).

8. D. M. Best and S. L. Rosen, *Polym. Eng. Sci.* **8**, 2, 116 (1968).

9. A. B. Metzner and J. C. Reed, *Am. Inst. Chem. Eng. J.*, **1**, 434 (1955).

10. G. K. Patterson, J. L. Zakin, and J. M. Rodriquez, *Ind. Eng. Chem.*, **61**, 1, 22 (1969).

11. J. W. Hoyt, *Trans. ASME J. Basic Eng.*, **94D**, 258 (1972).

12. J. L. Lumley, *Macromol. Rev.*, **7**, 263 (1973).

13. C. B. Wang, *Ind. Eng. Chem. Fundam.*, **11**, 546 (1972).

CHAPTER 17

Introduction to Continuum Mechanics

17.1 THREE-DIMENSIONAL STRESS AND STRAIN

The stress existing at any point in a material may always be resolved into components acting on the faces of a differential element in three arbitrary coordinate directions (Fig. 17.1). The stress components acting on the faces of the element are of two types, *normal stresses* (forces *normal* to the surface per unit surface area) and shear stresses (forces *parallel* to the surface per unit surface area). The first subscript conventionally identifies the direction perpendicular to the surface in question, and the second, the direction of the force itself. In general, there are three *normal* stresses, τ_{11}, τ_{22}, and τ_{33}, and six *shear* stresses, τ_{12}, τ_{13}, τ_{21}, τ_{23}, τ_{31}, and τ_{32}. These nine quantities, necessary for specifying the state of stress at a point completely, are the components of the *stress tensor* . They are conveniently written in matrix form:

$$\boldsymbol{\tau} = \begin{vmatrix} \tau_{11} & \tau_{12} & \tau_{13} \\ \tau_{21} & \tau_{22} & \tau_{23} \\ \tau_{31} & \tau_{32} & \tau_{33} \end{vmatrix} \tag{17.1}$$

The stress tensor is usually broken into an isotropic or hydrostatic pressure and a *deviatoric stress tensor*:

$$\boldsymbol{\tau} = - \begin{vmatrix} P & 0 & 0 \\ 0 & P & 0 \\ 0 & 0 & P \end{vmatrix} + \begin{vmatrix} p_{11} & \tau_{12} & \tau_{13} \\ \tau_{21} & p_{22} & \tau_{23} \\ \tau_{31} & \tau_{32} & p_{33} \end{vmatrix} \tag{17.2}$$

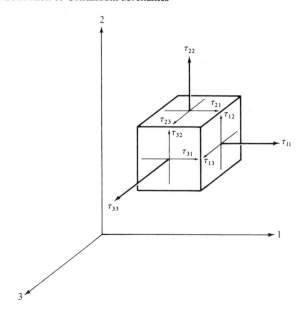

Figure 17.1   Stress components acting on an element of material. The equal-and-opposite stresses (necessary for equilibrium) on the hidden faces are not shown.

where P, the hydrostatic pressure is rather arbitrarily defined as

$$P = -\frac{(\tau_{11} + \tau_{22} + \tau_{33})}{3} \qquad (17.3)$$

and $p_{ii} = \tau_{ii} + P$ are the *deviatoric normal stresses*.

Caution: (a) The negative signs in (17.2) and (17.3) arise from the historical conventions of treating both hydrostatic pressure (an inward force on the element) and *tensile* stress (an outward force on the element) as positive. This convention is not adhered to universally, however.

Caution: (b) There are about as many different forms of notation as there have been books and papers written in this area. Be sure to understand the notation before plowing through any of the literature — that's half the battle.

17.2   GENERAL CONSTITUTIVE RELATIONS

The three-dimensional rate of strain at the point is expressed by a rate-of-strain tensor analogous to (17.1). Given a general *constitutive relation* between the stress and rate-of-strain tensors for a particular material, the mechanical response of that material can be described completely (in principle) under any circumstances. The goal of continuum mechanics is to develop such general constitutive relations and use them for predicting material response in the widest variety of situations. To complete the picture ultimately, it is the goal of molecular mechanics to predict these constitutive relations from the molecular structure of the material.[6] Unfortunately, from the standpoint of engineering application, the attainment of these goals is a long way off.

17.3   VISCOMETRIC FLOWS[1]

For *equilibrium viscometric* flows of *incompressible* fluids (the latter is a reasonable assumption for polymer melts and solutions in most engineering circumstances), the situation is brighter. With the coordinate directions assigned as described for viscometric flows in the last chapter, the shear stresses τ_{12} and τ_{21} are equal (or there would be a rotational flow component), and $\tau_{13} = \tau_{31} = 0$. There are two independent differences of the deviatoric normal stresses, commonly defined as

$$\sigma_1 = p_{11} - p_{22} \tag{17.4a}$$

$$\sigma_2 = p_{22} - p_{33} \tag{17.4b}$$

where σ_1 and σ_2 are the so-called first and second normal stress differences. Furthermore, there is only one nonzero component of the rate-of-strain tensor, $\dot{\gamma}_{12} = \dot{\gamma}$. Thus a complete description of an incompressible material in an equilibrium viscometric deformation requires only a knowledge of the three functions

$$\tau = \tau(\dot{\gamma}) \qquad \text{(the flow curve)} \tag{17.5a}$$

$$\sigma_1 = \sigma_1(\dot{\gamma}) \tag{17.5b}$$

$$\sigma_2 = \sigma_2(\dot{\gamma}) \tag{17.5c}$$

that is, the dependence of shear stress and first and second normal stress differences on the shear rate. Techniques for measuring functions (17.5b) and (17.5c) have been comprehensively reviewed.[1,5,7]

In nonequilibrium deformations, the response of the material depends not only on its present rate of strain, but on its complete strain history. Elaborate and very general constitutive relations have been developed to handle these situations, but they have achieved little engineering application because of their unavoidable complexity.

Example 1.   Consider the classical example of a nonviscometric flow, *simple elongation.* This situation arises, for example, if a weight is suspended from a rod of material. Note that here the velocity gradient *is not* perpendicular to the direction of fluid motion. Discuss the nature of the stress and rate of strain tensors.

Solution. Neglecting such things as surface tension forces on the sides of the rod, the deviatoric stress tensor has only one component, p_{11}, where the subscript 1 represents the axial direction. This will obviously produce a tensile elongation; so there will be a rate of tensile elongation component of the rate-of-strain tensor $\dot{\epsilon}_{11}$. For an *incompressible* material (Poisson's ratio = 0.5), the lateral dimension of the rod will contract to maintain the volume constant as the rod is extended, and the lateral *contractile* strains will be one-half the axial extension strain. Thus $\dot{\epsilon}_{11} = -2\dot{\epsilon}_{22} = -2\dot{\epsilon}_{33}$. There is no shearing; so all the shear strains are zero. Therefore, to describe this sort of deformation, a new material function η_e, the elongational or Trouton viscosity, is needed to relate tensile stress to the rate of tensile strain; that is, $\tau_{11} = \eta_e(\dot{\epsilon}_{11})\dot{\epsilon}_{11}$, or, equivalently, the function $\tau_{11} = \tau_{11}(\dot{\epsilon}_{11})$. For *incompressible Newtonian fluids*, it may be shown that $\eta_e = 3\eta$.

Converging and diverging flows (in which the fluid passes from a large channel into a smaller one, and vice versa) are examples of nonviscometric flows that combine shearing and elongational deformations. Such flows are frequently encountered in polymer processing operations. Nevertheless, even for Newtonian fluids, exact solutions have not been obtained, although useful approximations and numerical results are available. The treatment of such flows for viscoelastic fluids is still in its infancy.

17.4 INTERPRETATION OF THE NORMAL STRESSES

The presence of deviatoric normal stress components in viscometric flow implies fluid elasticity, and these components are physically real quantities. The well-known *Weissenberg effect* is a good example. When a vertical rotating rod is immersed in a container of inelastic fluid, the fluid is flung outward by centrifugal force, and its surface assumes a parabolic profile, with the lowest point at the rod A viscoelastic fluid, on the other hand, will actually climb up the rod. This is a viscometric Couette flow, as described in the previous chapter. The behaviour of the viscoelastic fluid can be rationalized by imagining it to consist of rubber bands, stretched by the rotation along a fluid streamline in the tangential (θ or 1) direction, at constant radius. The rubber bands are obviously in tension, and hence $p_{11} = p_{\theta\theta}$ is positive (a "hoop" stress, as in the bands of a barrel). This positive hoop stress places the cylinder of material within under a compressive stress in the radial direction (think of what wrapping a rubber band around your finger does); that is, a negative $p_{22} = p_{rr}$. Since the inwardly compressed material is prevented from moving inward by the presence of the rod,

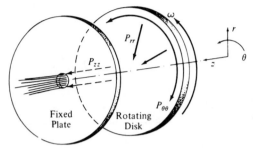

Figure 17.2   Normal stresses in the elastic melt extruder.[8]

it has nowhere to go but up the rod under the influence of a compressive axial stress; a negative $p_{33} = p_{zz}$. This does not mean to imply that centrifugal force is no longer acting; merely that it is overwhelmed by the elastic normal stresses.

Similar reasoning leads to the conclusion that in cone-and-plate or disk–plate viscometers (Section 16.7) there will be forces tending to push the cone and plate or the disk and plate apart. Indeed, there are, and the measurement of these forces provides a means of determining the functions $\sigma_1(\dot{\gamma})$ and $\sigma_2(\dot{\gamma})$.[1,5,7] Also, for the case of torsional flow, Maxwell and Scalora[8] show that if a hole is drilled through the plate along the axis of rotation, a screwless extruder is obtained, as the axial normal stress pushes a polymer melt through the hole (Fig. 17.2). Unlike a screw extruder, however, this device only works with viscoelastic fluids.

REFERENCES

1.   B. D. Coleman, H. Markovitz, and W. Noll, *Viscometric Flow of Non-Newtonian Fluids*, Springer-Verlag, New York, 1966.

2.   S. Middleman, *The Flow of High Polymers*, Wiley, New York, 1968.

3.   R. M. Christensen, *Theory of Viscoelasticity. An Introduction*, Academic Press, New York, 1971.

4.   R. Darby, *Viscoelastic Fluids. An Introduction to their Properties and Behavior*, Marcel Dekker, New York, 1976.

5.   R. B. Bird, R. C. Armstrong, and O. A. Hassager, *Dynamics of Polymeric Liquids*, Vol. I: Fluid Mechanics, Wiley, New York, 1977.

6.   R. B. Bird, O. A. Hassager, R. C. Armstrong, and C. F. Curtiss, *Dynamics of Polymeric Liquids*. Vol. II: Kinetic Theory, Wiley, New York, 1977.

7.   K. Walters, *Rheometry*, Halsted Press, New York, 1975.

8.   B. Maxwell, and A. J. Scalora, *Mod. Plast.*, 37, 107 (1959).

CHAPTER 18

Linear Viscoelasticity

18.1 INTRODUCTION

Engineers have traditionally been concerned with two separate and distinct classes of materials: the viscous fluid and the elastic solid. Elaborate design procedures have been based on these concepts, and they have worked pretty well because most traditional materials (water, air, steel, concrete), at least to a good approximation, fall into one of these categories. The realization has grown, however, that these categories represent only the extremes of a broad spectrum of material response, and polymers fall somewhere in between, giving rise to some of the unusual properties of melts and solutions described previously. Other examples are important in the structural applications of polymers. In a common engineering stress–strain test, a sample is strained at an approximately constant rate, and the stress is measured as a function of strain. With traditional solids, the stress–strain curve is pretty much independent of the rate at which the material is strained. The stress–strain properties of many polymers are markedly *rate dependent*, however. Similarly, polymers often exhibit pronounced creep and stress relaxation (to be defined shortly). While other materials also exhibit such behaviour (metals near their melting points, for example), at normal temperatures it is negligible and not usually included in design calculations. If the time-dependent behaviour of polymers is ignored, the results can sometimes be disastrous.

18.2 MECHANICAL MODELS FOR LINEAR VISCOELASTIC RESPONSE

As an aid in visualizing viscoelastic response, we introduce two *linear* mechanical models to represent the extremes of the mechanical response spectrum. The

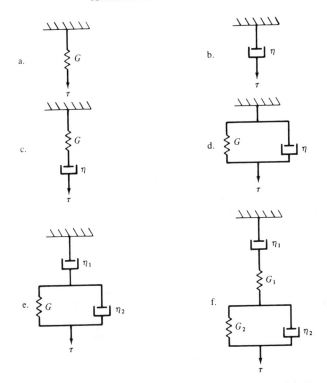

Figure 18.1 Linear viscoelastic models. (*a*) Linear elastic; (*b*) linear viscous; (*c*) Maxwell element; (*d*) Voigt–Kelvin element; (*e*) three parameter; (*f*) four parameter.

spring in Fig. 18.1*a* represents a *linear elastic* or Hookean *solid* whose constitutive equation (relation of stress to strain and time) is simply $\tau = G\gamma$, where G is a (constant) shear modulus. Similarly, a *linear viscous* or Newtonian *fluid* is represented by a dashpot (some sort of piston moving in a cylinder of Newtonian fluid) whose constitutive equation is $\tau = \eta\dot{\gamma}$, where η is a (constant) viscosity. The strain is represented by the extension (stretching) of the model. Although the models developed here are to be visualized in tension, the notation used is for pure shear (viscometric) deformation. The equations are equally applicable to tensile deformation by replacing the shear stress τ with the tensile stress σ, the shear strain γ with the tensile strain ϵ, Hooke's modulus G with Young's (tensile) modulus E, and the Newtonian (shear) viscosity η with Trouton's (tensile) viscosity η_e.

Some authorities object strongly to the use of mechanical models to represent materials. They point out that materials are not made up of springs and dashpots. True, but neither are they made up of mathematical equations, and it's a lot

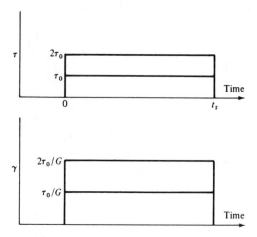

Figure 18.2   Response of spring.

easier for most people to visualize the deformation of springs and dashpots than the solutions to equations.

A word is needed about the meaning of the term *linear*. For the present, a linear response is defined as one in which the *ratio* of overall stress to overall strain, that is, the overall modulus, is a function of *time only*, not of the magnitudes of stress or strain,

$$G \equiv \frac{\tau}{\gamma} = G \quad (t \text{ only}) \qquad \text{for linear response} \qquad (18.1)$$

The Hookean spring responds instantaneously to reach an equilibrium strain γ upon application of a stress, and the strain remains constant as long as the stress is maintained constant. Sudden removal of the stress results in instantaneous recovery of the strain (Fig. 18.2). Doubling the stress on the spring simply doubles the resulting strain; so the spring is linear according to (18.1). (In assuming that the spring instantaneously reaches an equilibrium strain under the action of a suddenly applied constant stress, we have neglected inertial effects. Although it is not necessary to do so, including them would contribute little to the present discussion.) If a constant stress is suddenly applied to the dashpot, the strain increases with time according to $\gamma = (\tau/\eta)t$ (considering the strain to be zero when the stress is applied; Fig. 18.3). Doubling the stress doubles the slope of the strain–time line, and at any time, the modulus $G \equiv \tau/\gamma = \eta/t = G$ (t only). So the dashpot is also linear.

It may be shown that *any combination of linear elements must be linear*; so any models based on these linear elements, no matter how complex, can only represent linear response. Just how realistic is linear response? Well, for most polymers at strains greater than a percent or so (or rates of strain greater than

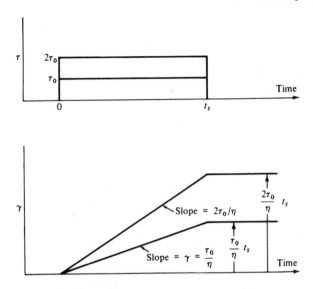

Figure 18.3   Response of dashpot.

$0.1\,\mathrm{sec}^{-1}$), it's not a good quantitative description. Moreover, even within the *limit of linear viscoelasticity*, it usually requires a large number of linear elements (springs and dashpots) to provide an accurate quantitative description of response. Hence the quantitative applicability of simple linear models (i.e., those with a few springs and dashpots) is limited, but they are extremely valuable in visualizing viscoelastic response, and in understanding how and why variations in molecular structure influence that response.

A.   The Maxwell Element

James Clerk Maxwell realized that neither a linear viscous element (dashpot) nor a linear elastic element (spring) was sufficient to describe his experiments on the deformation of asphalts, and so he proposed a simple series combination of the two, the *Maxwell element* (Fig. 18.1c). In a Maxwell element, the spring and dashpot support the same stress; so

$$\tau = \tau_{\mathrm{spring}} = \tau_{\mathrm{dashpot}} \qquad (18.2)$$

Furthermore, the overall strain of the element is the sum of the strains in the spring and dashpot:

$$\gamma = \gamma_{\mathrm{spring}} + \gamma_{\mathrm{dashpot}} \qquad (18.3)$$

Differentiating (18.3) with respect to time,

$$\dot{\gamma} = \dot{\gamma}_{\mathrm{spring}} + \dot{\gamma}_{\mathrm{dashpot}} \qquad (18.4)$$

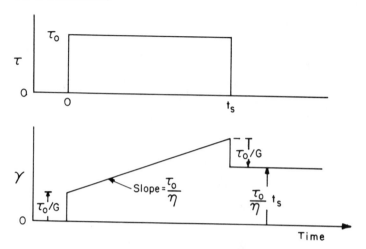

Figure 18.4   Creep response of Maxwell element.

Realizing that $\dot\gamma_{\text{dashpot}} = \tau/\eta$ and $\dot\gamma_{\text{spring}} = \dot\tau/G$, plugging in and rearranging gives the differential equation for the Maxwell element:

$$\tau = \eta\dot\gamma - \left(\frac{\eta}{G}\right)\dot\tau = \eta\dot\gamma - \lambda\dot\tau \qquad (18.5)$$

The quantity $\lambda = \eta/G$ has the dimensions of time and is known as a *relaxation time*. Its physical significance will be apparent shortly.

Creep Testing

Let's examine the response of the Maxwell element in two mechanical tests commonly applied to polymers. First consider a *creep* test, in which a *constant stress* is instantaneously (or at least very rapidly) *applied* to the material, and the resulting *strain is followed as a function of time*. Deformation after removal of the stress is known as *creep recovery*. As shown in Fig. 18.4, the sudden application of stress to a Maxwell element causes an instantaneous stretching of the spring to an equilibrium value of τ_0/G, where τ_0 is the (constant) applied stress. The dashpot extends linearly with time with a slope of τ_0/η, and will continue to do so as long as the stress is maintained. Thus *the Maxwell element is a fluid* because it will continue to deform as long as it is stressed. When the stress is released, the spring immediately contracts by an amount equal to its original extension, a process known as *elastic recovery*. The dashpot, of course, does not recover, leaving a *permanent set* of $(\tau_0/\eta)t_s$, the amount the dashpot has extended during the application of stress.

Although real materials never show sharp breaks in a creep test, as does the

Maxwell element, the Maxwell element does exhibit the phenomena of elastic strain, creep, recovery, and permanent set, which are often observed with real materials.

Stress Relaxation

Another important test used to study viscoelastic response is *stress relaxation.* A stress-relaxation test consists of suddenly *applying a strain* to the sample and *following the stress as a function of time as the strain is held constant.* When the Maxwell element is strained instantaneously, only the spring can respond initially (for an infinite rate of strain, the resisting force in the dashpot is infinite) to a stress of $G\gamma_0$, where γ_0 is the constant applied strain. The extended spring then begins to contract, but the contraction is resisted by the dashpot. The more the spring retracts, the smaller its restoring force becomes, and the rate of retraction drops correspondingly. Solution of the differential equation with $\dot{\gamma} = 0$ and the boundary condition $\tau = G\gamma_0$ at $t = 0$ shows that the stress undergoes a first-order exponential decay,

$$\tau = G\gamma_0 e^{-t/\lambda} \tag{18.6}$$

Thus, the relaxation time λ is the time constant for the exponential decay, that is, the time required for the stress to decay to a factor of $1/e$ or 37% of its initial value. The stress asymptotically drops to zero as the spring approaches complete retraction (Fig. 18.5).

Stress-relaxation data for linear polymers actually look like the curve for the

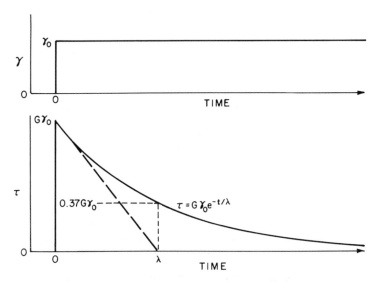

Figure 18.5   Stress relaxation of Maxwell element.

Maxwell element. Unfortunately, they can't often be fitted quantitatively with a single value of G and a single value of λ. The decay is not really first order.

Example 1.   Examine the response of a Maxwell element in an engineering stress–strain test, a test in which the rate of tensile strain is maintained (approximately) constant at $\dot{\epsilon}_0$.

Solution.   Rewriting (18.5) in tensile notation

$$\sigma = \eta_e \dot{\epsilon}_0 - \left(\frac{\eta_e}{E}\right)\dot{\sigma} \quad \text{or} \quad \sigma + \left(\frac{\eta_e}{E}\right)\frac{d\sigma}{dt} = \eta_e \dot{\epsilon}_0 = \text{constant}$$

The solution to this differential equation with the boundary condition $\sigma = 0$ at $t = 0$ is

$$\sigma = \eta_e \dot{\epsilon}_0 \left\{ 1 - \exp\left[-\left(\frac{E}{\eta_e}\right)t \right] \right\}$$

Since $d\epsilon/dt = \dot{\epsilon}_0 = \text{constant}$, and $\epsilon = 0$ when $t = 0$,

$$\epsilon = \dot{\epsilon}_0 t$$

The stress–strain curve is then

$$\sigma = \eta_e \dot{\epsilon}_0 \left\{ 1 - \exp\left[-\left(\frac{E}{\eta_e \dot{\epsilon}_0}\right)\epsilon \right] \right\}$$

This response is sketched in Fig. (18.6). Note that at any given strain, σ increases with the *rate of strain* $\dot{\epsilon}_0$; that is, the material appears "stiffer" (has a higher

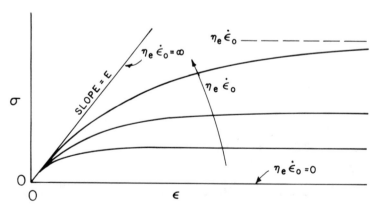

Figure 18.6   Response of a Maxwell element under constant rate of tensile strain (Example 1).

modulus). As crude as this model might prove to be in fitting actual experimental data, it does account, at least qualitatively, for many of the observed properties of linear polymers in engineering stress–strain tests.

B.   The Voigt–Kelvin Element

Well, if a series combination of a spring and dashpot has its drawbacks, the next logical thing to try is a parallel combination, a Voigt or Voigt–Kelvin element, (Fig. 18.1d). Here it is assumed that the crossbars supporting the spring and dashpot always remain parallel so that *the strain in each element is the same*:

$$\gamma = \gamma_{\text{spring}} = \gamma_{\text{dashpot}} \tag{18.7}$$

The stress supported by the element is the sum of the stresses in the spring and the dashpot:

$$\tau = \tau_{\text{spring}} + \tau_{\text{dashpot}} \tag{18.8}$$

Combination of (18.7) and (18.8) with the equations for the deformation of the spring and dashpot gives the differential equation for the Voigt–Kelvin element:

$$\tau = \eta\dot{\gamma} + G\gamma \tag{18.9}$$

When the stress is suddenly applied in a creep test, only the dashpot offers an initial resistance to deformation; so the initial slope of the strain versus time curve is τ_0/η. As the element is extended, the spring provides an increasingly greater resistance to further extension, and so the rate of creep decreases. Eventually, the system comes to equilibrium with the spring alone supporting the stress (with the rate of strain zero, the resistance of the dashpot is zero). The equilibrium strain is simply τ_0/G. Quantitatively, the response is an exponential rise:

$$\gamma = \frac{\tau_0}{G}[1 - e^{-t/\lambda}] \tag{18.10}$$

If the stress is removed after equilibrium has been reached, the strain decays exponentially:

$$\gamma = \frac{\tau_0}{G}e^{-t/\lambda} \tag{18.11}$$

Note that the Voigt–Kelvin model does not continue to deform as long as stress is applied, and it does not exhibit any permanent set (Fig. 18.7). It therefore represents a viscoelastic *solid* and gives a fair qualitative picture of the creep response of some crosslinked polymers.

The Voigt–Kelvin model is not suited for representing stress relaxation. The application of an instantaneous strain would be met by an infinite resistance in the dashpot, and so would require the application of an infinite stress, which is obviously unrealistic.

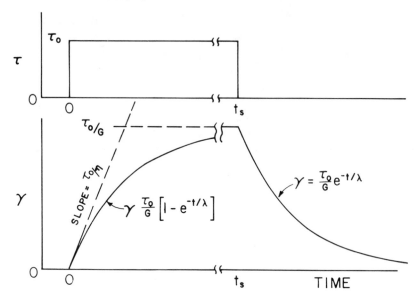

Figure 18.7   Creep response of a Voigt–Kelvin element.

C.   The Three-Parameter Model

The next step in the development of linear viscoelastic models is the so-called three-parameter model (Fig. 18.1*e*). The differential equation for this model may be written in operator form as

$$\left[1 + \lambda_1 \frac{d}{dt}\right]\tau = \eta_1\left[1 + \lambda_2\frac{d}{dt}\right]\dot{\gamma} \tag{18.12}$$

where $\lambda_1 = (\eta_1 + \eta_2)/G$ and $\lambda_2 = \eta_2/G$. Further, the form of (18.12) suggests modification by adding higher order derivatives and more constants:

$$\left[1 + \lambda_1\frac{d}{dt} + \xi_1\frac{d^2}{dt^2} + \cdots\right]\tau = \eta_1\left[1 + \lambda_2\frac{d}{dt} + \xi_2\frac{d^2}{dt^2} + \cdots\right]\dot{\gamma} \tag{18.13}$$

This, of course, will fit data to any desired degree of accuracy if enough terms are used.

18.3   THE FOUR-PARAMETER MODEL AND MOLECULAR RESPONSE

The four-parameter model (Fig. 18.1*f*) is a series combination of a Maxwell element with a Voigt–Kelvin element, and its creep response is the sum of

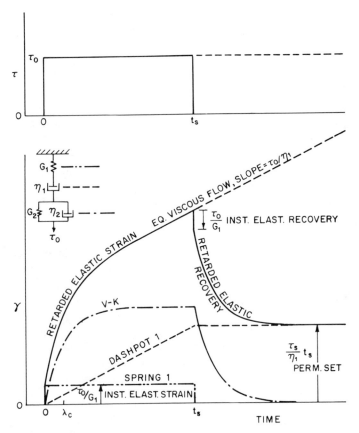

Figure 18.8   Creep response of a four-parameter model.

their individual responses, as summarized in Fig. 18.8. It provides at least a qualitative representation of all the phenomena generally observed with visco-elastic materials: instantaneous elastic strain, retarded elastic strain, equilibrium viscous flow, instantaneous elastic recovery, retarded elastic recovery, and permanent set. Of equal importance is the fact that the model parameters can be identified with the various molecular response mechanisms in polymers and the model can therefore be used to predict the influences that changes in molecular structure will have on mechanical properties. The following analogies may be drawn:

1.   Dashpot 1 represents molecular slippage. This slip of polymer molecules past one another is responsible for flow. The value of η_1 alone (molecular friction in slip) governs the equilibrium flow of the material.

2.   Spring 1 represents the elastic straining of bond angles and lengths. All bonds in polymer chains have equilibrium angles and lengths. The value of G_1 characterizes the resistance to deformation from these equilibrium values. Since these deformations involve interatomic bonding, they occur essentially instantaneously from a macroscopic point of view. This type of elasticity is known thermodynamically as *energy elasticity*, because the straining of bond angles and lengths increases the material's internal energy (see Chapter 14).

3.   Dashpot 2 represents the resistance of the polymer chains to uncoiling and coiling, caused by temporary mechanical entanglements of the chains and molecular friction during these processes. Since coiling and uncoiling require cooperative motion of many chain segments, they cannot occur instantaneously and hence account for retarded elasticity.

4.   Spring 2 represents the restoring force brought about by the thermal agitation of the chain segments, which tends to return chains oriented by a stress to their most random or highest entropy configuration. This type of restoring force is known, therefore, as *entropy elasticity* (Chapter 14).

The magnitude of the time scale of Fig. 18.8 will of course depend on the particular material and its temperature, that is, on the values of the model parameters. Well below T_g, for example, where η_1 and η_2 are very large, t_s might be on the order of days or weeks in order that appreciable retarded elasticity and flow be observed. Above T_g, t_s might be only seconds or less to permit the deformation shown. An important thing to keep in mind is that designs based on short-term property measurements will be inadequate if the object must support a stress for longer periods of time.

Example 2.   Using the four-parameter model as a basis, sketch qualitatively the effects of (a) increasing molecular weight and (b) increasing degrees of cross-linking on the creep response of a linear amorphous polymer.

Solution

a.   As discussed in Chapter 15, the equilibrium zero-shear (linear) viscosity of polymers, represented by η_1 in the model, increases with the 3.4 power of $\bar{M}_w$. Thus, the slope in the equilibrium flow region τ_0/η_1 is greatly decreased as the molecular weight increases, and the permanent set $(\tau_0/\eta_1)t_s$, is reduced correspondingly (Fig. 18.9).

b.   Light crosslinking represents the limit of case (a) above when the molecular weight reaches infinity, as all the chains are hooked together by crosslinks. Under these conditions they *can't* slip past one another; therefore η_1 becomes infinite. If the crosslinking is light (the crosslinks few and far between), as in a rubber band, coiling and uncoiling won't be hindered appreciably. Note that crosslinking converts the material from a fluid to a

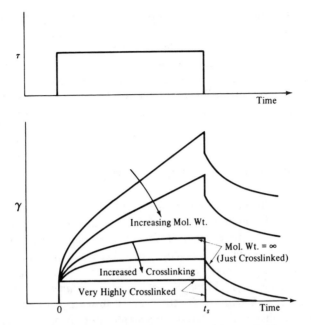

Figure 18.9 The effects of molecular weight and crosslinking on the creep response of an amorphous polymer.

solid (it eventually reaches an equilibrium strain under the application of a constant stress), and it eliminates permanent set. The equilibrium modulus will be on the order of 10^6 to 10^7 dynes/cm^2 (10^5-10^6 N/m^2), the characteristic "rubbery" modulus.

Further crosslinking begins to hinder the ability of the chains to uncoil and raises the restoring force (increases η_2 and G_2). At high degrees of crosslinking, as in hard rubber (ebonite), the only response mechanism left is straining bond angles and lengths, giving rise to an almost perfectly elastic material with a modulus on the order of 10^{10} to 10^{11} dynes/cm^2 (10^9-10^{10} N/m^2), the characteristic "glassy" modulus.

The four-parameter model nicely accounts for the interesting examples of viscoelastic response mentioned earlier. For example, dashpot 1 allows viscous flow, while the elastic restoring forces of springs 1 and 2 provide the "rubber band" elasticity responsible for the Weissenberg effect. In engineering stress–strain tests, the moduli of polymers are observed to increase with the applied rate of strain. At high rates of strain, spring 1 provides the major response mechanism. As the rate of strain is lowered, dashpot 1 and the Voigt–Kelvin element contribute more and more to the overall deformation, giving a greater strain at any stress, that is, a lower modulus. When "silly putty" is bounced (stress applied rapidly for a short period of time), spring 1 again provides the major response mechanism. There is no time for appreciable flow of the dashpots 1 and 2; so not much of the initial potential energy is converted to heat through the molecular friction involved in slippage and uncoiling, and the material behaves in an almost perfectly elastic fashion. When it's stuck on the wall, the stress — in this case due to its own weight — is applied for a long period of time, and it flows downward as a result of the molecular slip represented by dashpot 1.

18.4   VISCOUS OR ELASTIC RESPONSE? THE DEBORAH NUMBER[1]

Thus, whether a viscoelastic fluid behaves as an elastic solid or a viscous liquid depends on the relation between the time scale of the deformation to which it is subjected and the time required for the time-dependent mechanisms to respond. Strictly speaking, the concept of a single relaxation time applies only to first-order response and therefore is not applicable to real materials, in general. Nevertheless, *a characteristic time,* λ_c, *for any material can always be defined* as, for example, the time required for the material to reach $1 - (1/e)$ or 63.2% of its ultimate retarded elastic response to a step change. A precise value is rarely necessary. It's simply a means of characterizing the magnitude of the rate of a material's time-dependent elastic response, short λ_c's indicating rapid response and large λ_c's sluggish response. The ratio of this characteristic time to the time scale of the deformation is the *Deborah number*:

$$\text{De} = \frac{\lambda_c}{t_s} \qquad (18.14)$$

At high Deborah numbers ($\text{De} \gg 1$), response will appear elastic, and at low Deborah numbers ($\text{De} \to 0$), viscous.*

Consider, for example, the creep response of the four-parameter model (Fig. 18.8). For this model, a logical choice for λ_c would be the time constant for its Voigt–Kelvin component η_2/G_2. For $\text{De} \gg 1$ ($t_s \ll \lambda_c$), the Voigt–Kelvin element and dashpot 1 will be essentially immobile, and the response will be due almost entirely to spring 1, that is, purely elastic. For $\text{De} \to 0$ ($t_s \gg \lambda_c$), the instantaneous and retarded elastic response mechanisms have reached equilibrium; so the only remaining response will be the purely viscous flow of dashpot 1, and the deformation due to viscous flow will completely overshadow that due to the elastic response mechanisms (imagine the creep curve of Fig. 18.8 extended a meter or so beyond the page). Under conditions when $\text{De} \to 0$, materials can be treated by the techniques outlined in Chapter 16 for purely viscous fluids.

*"The Mountains flowed before the Lord" [Song of Deborah, Judges 5:5]. For the Lord, t_s is exceedingly large; see reference 1.

Modifications of the devices described in Chapter 16 can also be used to obtain information on the material's elastic properties. For example, if the stress is suddenly removed from a rotational viscometer, the creep recovery or elastic recoil of the material can be followed. This provides a value of λ_c for the material.

Example 3.   Thermocouples and Pitot tubes inserted in a flowing stream of a viscoelastic fluid often give erroneous results. Explain.

Solution.   When a viscoelastic fluid in equilibrium flow (De → 0) encounters a probe, it must make a sudden (De ≫ 1) jog to get around it. The retarded elastic response mechanisms simply cannot respond fast enough in the immediate vicinity of the probe, which for all practical purposes behaves as if it were covered with a solid plug. What it measures, therefore, is not characteristic of the fluid in an unobstructed stream.

Example 4.   (This is believed due to Prof. A. B. Metzner.) A paper cup containing water is placed on a stump. A .22-caliber bullet fired at the cup passes cleanly through, leaving the cup sitting on the stump. The water is replaced by a dilute polymer solution in a second cup. This time, the bullet knocks the cup 25 ft from the stump. Explain.

Solution.   The characteristic time for a low-molecular-weight fluid such as water is extremely short, much shorter than the time it takes the bullet to pass through the cup (t_s). This, then, is a low-De situation. The water behaves as a viscous fluid. The bullet transfers a little momentum to it through viscous friction, but not enough to dislodge the cup. Adding a polymeric solute raises the characteristic time many orders of magnitude, to the point where this becomes a high-De experiment (the polymer chains can't respond fast enough to get out of the way of the bullet). The bullet, in effect, slams into a solid and transfers most of its momentum to the bullet–fluid–cup system, carrying it beyond the stump.

18.5   QUANTITATIVE APPROACHES[2-5]

Although the four-parameter model is so useful from a conceptual standpoint, the four parameters usually are not sufficient to provide an accurate "fit" of most experimental data. To do so, and to infer some detailed information about molecular response, more general models have been developed. The *generalized Maxwell model* (Fig. 18.10), is used to describe stress-relaxation experiments. The stress relaxation of an individual Maxwell element is given by

$$\tau_i(t) = \gamma_0 G_i e^{-t/\lambda_i} \tag{18.15}$$

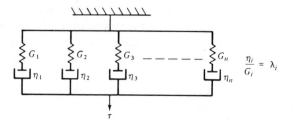

Figure 18.10   Generalized Maxwell model.

where $\lambda_i = \eta_i/G_i$. The relaxation of the generalized model, in which the individual elements are all subjected to the same constant strain γ_0, is then

$$\tau(t) = \gamma_0 \sum_{i=1}^{n} G_i e^{-t/\lambda_i} \tag{18.16}$$

Expressed in terms of a time-dependent modulus $G(t)$, the response is

$$G(t) = \frac{\tau(t)}{\gamma_0} = \sum_{i=1}^{n} G_i e^{-t/\lambda_i} \tag{18.17}$$

Now, if n is large, the summation of individual discrete moduli in (18.17) may be approximated by the integral of a *continuous distribution* of *relaxation times* $G(\lambda)$:

$$G(t) = \int_0^{\infty} G(\lambda) e^{-t/\lambda} d\lambda \tag{18.18}$$

Note that while the G_i have units of dynes/cm^2, $G(\lambda)$ is in dynes/cm^2 sec. Note also that if the generalized Maxwell model is to represent a viscoelastic solid, for example, a crosslinked polymer, at least one of the viscosities has to be infinite.

For creep tests, a generalized Voigt–Kelvin model is used (Fig. 18.11). The creep response of an individual Voigt–Kelvin element is given by

$$\gamma_i(t) = \tau_0 J_i(1 - e^{-t/\lambda_i}) \tag{18.19}$$

where $J_i = 1/G_i$ is a *compliance* (cm^2/dyne). The response of the array, in which each element is subjected to the same constant applied stress τ_0, is then

$$\gamma(t) = \tau_0 \sum_{i=1}^{n} J_i(1 - e^{-t/\lambda_i}) \tag{18.20}$$

Or, in terms of a time-dependent compliance $J(t)$,

$$J(t) = \frac{\gamma(t)}{\tau_0} = \sum_{i=1}^{n} J_i(1 - e^{-t/\lambda_i}) \tag{18.21}$$

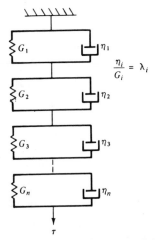

Figure 18.11   Generalized Voigt–Kelvin model.

Again, for large n, the discrete summation above may be approximated by

$$J(t) = \int_0^\infty J(\lambda)(1 - e^{-t/\lambda})d\lambda \qquad (18.22)$$

where $J(\lambda)$ is the *continuous distribution of retardation times* (cm^2/dyne sec).

If the generalized Voigt–Kelvin model is to represent a viscoelastic liquid, a linear polymer, for example, the modulus of one of the springs must be zero (infinite compliance), leaving a simple dashpot in series with all the other Voigt–Kelvin elements. Sometimes, the steady-flow response of this lone dashpot, $\gamma_{\text{dashpot}} = (\tau_0/\eta)t$, is subtracted from the overall response, leaving the compliances to represent only the elastic contributions to the overall response:

$$\gamma(t) = \frac{\tau_0}{\eta}t + \tau_0 \sum_{i=1}^n J_i^\dagger (1 - e^{-t/\lambda_i}) \qquad (18.23)$$

$$J(t) = \frac{\gamma(t)}{\tau_0} = \frac{t}{\eta} + \sum_{i=1}^n J_i^\dagger (1 - e^{-t/\lambda_i}) \qquad (18.24)$$

$$J^\dagger(t) = J(t) - \frac{t}{\eta} = \sum_{i=1}^n J_i^\dagger (1 - e^{-t/\lambda_i}) \qquad (18.25)$$

$$J^\dagger(t) = J(t) - \frac{t}{\eta} = \int_0^\infty J^\dagger(\lambda)(1 - e^{-t/\lambda})d\lambda \qquad (18.26)$$

Here, the daggers indicate that the purely viscous flow has been removed and is treated separately.

Application of the discrete equations (18.16), (18.17), (18.20), and (18.21) generally involves a large n to describe the data accurately, thus requiring an impractically large number of parameters λ_i and $G_i = 1/J_i$. It has been suggested in reference 6, however, that the parameters are related by

$$\lambda_i = \frac{\lambda_0}{i^\alpha} \tag{18.27}$$

and

$$G_i = \frac{1}{J_i} = \frac{\eta_0}{\sum\limits_{i=1}^{n} \lambda_i} \tag{18.28}$$

Equations (18.27) and (18.28) require that the G_i all be the same and that $\eta_0 = \sum\limits_{i} \eta_i$ (with $\eta_i = \lambda_i G_i$). They reduce the number of necessary parameters to three: η_0, the *steady-state*, zero-shear viscosity; λ_0, a maximum relaxation time; and α, an empirical constant. Molecular theories[7] for dilute polymer solutions predict $\alpha = 2$, but for concentrated solutions and melts, better fits are obtained with α's between 2 and 4.[6]

Procedures are available for extracting the continuous distributions $G(\lambda)$ or $J(\lambda)$ from experimental data.[2,4] Once known, they can be used in (18.18) or (18.22) for predicting creep or stress relaxation response in the linear region. Of what use is that, if you first had to measure the response to determine these functions? Well, for one thing, $G(\lambda)$ can in principle be obtained from $J(\lambda)$, and vice versa, if one distribution is known over the range $0 < \lambda < \infty$. Although the functions are never obtainable over the complete range, reasonable approximations are available. Thus, creep response can be predicted from stress-relaxation measurements, and vice versa. This interconvertability also applies to a variety of linear mechanical responses in addition to the two types discussed here, as illustrated in the next section. Furthermore, the shape of the distributions provides the polymer scientist with information on molecular response mechanisms within the polymer. For example, peaks in a certain region of λ might imply motion of side chains on the molecules. This type of information can lead to the "design" of polymers with the type of side chains needed to provide particular mechanical properties.

18.6   THE BOLTZMANN SUPERPOSITION PRINCIPLE

Suppose a material initially free of stress and strain is subjected to a test in which a strain $\gamma(t_0)$ is suddenly imposed at $t = 0$ and maintained constant for a while. This is classical stress relaxation, and the stress will decay according to the material's time-dependent relaxation modulus, $G(t)$; that is, $\tau(t) = G(t)\gamma(t_0)$. Now, however, at time t_1, the strain is suddenly changed to a new level $\gamma(t_1)$,

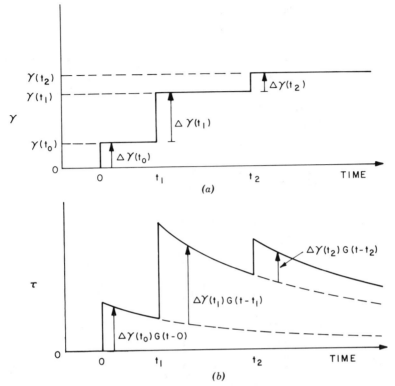

Figure 18.12 Boltzmann superposition principle. (*a*) Applied strain history; (*b*) resulting stress history.

held there for a while, and then at t_2 changed to $\gamma(t_2)$, and so on, as sketched in Fig. 18.12*a*. What happens to the stress as a result of this strain history? Well, way back in 1876, Boltzmann suggested that the stresses resulting from each individual strain increment should be linearly additive:

$$\tau(t) = \sum_{i=0}^{n} \Delta\gamma(t_i)G(t-t_i) \qquad (\text{for } t > t_i) \qquad (18.29)$$

where $\Delta\gamma(t_i) = \gamma(t_i) - \gamma(t_{i-1})$. This behavior is sketched in Fig. 18.12*b*. According to Boltzmann, the stress in the material at any time t depends on its *entire past strain history*, although since $G(t)$ decreases with time, the further back a $\Delta\gamma(t_i)$ has occurred, the smaller will be its influence in the present. This leads to the anthropomorphic concept of viscoelastic materials having a memory, a fading memory, at that, with $G(t)$ sometimes known as the *memory function*. (The concept, of course, is valid even in the absence of linear additivity — it's just much more difficult to quantify.) When all is said and done, probably the best definition of a linear material is simply one that follows Boltzmann's principle. Thus, spring–dashpot models, which are linear, automatically follow Boltzmann's principle.

Example 5.   A Maxwell element is initially free of stress and strain. At time $t = 0$, a strain of magnitude γ_0 is suddenly applied and maintained constant until $t = \lambda/2$, at which time the strain is suddenly reversed to a value of $-\gamma_0$ and maintained at that value (Fig. 18.13a). Obtain an expression for $\tau(t)$ and plot the result.

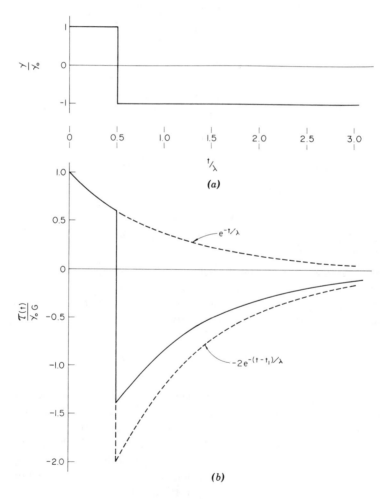

Figure 18.13   Response of a Maxwell element (Example 5).

Solution.   From (18.6) for a Maxwell element

$$G(t) \equiv \frac{\tau(t)}{\gamma_0} = Ge^{-t/\lambda}$$

For this particular strain history, $\Delta\gamma(t_0) = +\gamma_0$ and $\Delta\gamma(t_1) = -2\gamma_0$. Plugging these into (18.29) gives

$$\tau(t) = \Delta\gamma(t_0)G(t-0) + \Delta\gamma(t_1)G(t-t_1)$$
$$= \gamma_0 Ge^{-t/\lambda} - 2\gamma_0 Ge^{-(t-t_1)/\lambda}$$
$$= \gamma_0 G[e^{-t/\lambda} - 2e^{-(t-t_1)/\lambda}]$$

This result is plotted in dimensionless form in Fig. 18.13b. Keep in mind that the second term only has physical significance for $t > t_1 = \lambda/2$.

Of course, not all strain histories consist of a nice series of finite step changes. No matter how an applied strain varies with time, however, it can always be approximated by a series of differential step changes, for which (18.29) becomes

$$\tau(t) = \int_{\gamma(-\infty)}^{\gamma(t)} G(t-t')d\gamma(t') = \int_{-\infty}^{t} G(t-t')\frac{d\gamma(t')}{dt'}dt'$$

$$= \int_{-\infty}^{t} G(t-t')\dot{\gamma}(t')dt' \qquad (18.30)$$

where $t = present$ time and $t' = past$ time. A word is needed about the somewhat bizarre (but fairly standard) notation in (18.30). The implication is that to evaluate the stress at the present time t, we must integrate over the *entire past strain history* of the sample; hence the lower limit of $t' = -\infty$. In some (but not all) cases, it is convenient to assume that $\tau = \gamma = 0$ for $t' < 0$, in which case the lower limit on the integrals becomes zero. Equation (18.30) allows calculation of $\tau(t)$ from stress-relaxation data $G(t)$, for *any* applied strain sequence as long as response is linear. Furthermore, by inversion of (18.30), it is possible (at least in principle) to obtain $G(t)$ from *any* test in which $\tau(t)$ and $\gamma(t)$ are measured. Analogs of (18.29) and (18.30) may be written in terms of the creep compliance $J(t)$, when the independent variable is $\tau(t)$ and it is desired to calculate $\gamma(t)$.

Example 6.   Solve Example 1 by applying the Boltzmann superposition principle, thereby demonstrating how stress-time response in an engineering stress–strain test may be predicted from stress-relaxation data.

Solution.   The *tensile* stress-relaxation modulus for a Maxwell element is

$$E_r(t) \equiv \frac{\sigma(t)}{\epsilon_0} = Ee^{-t/\lambda}$$

Equation (18.30) becomes, in tensile notation, with the assumption that $\sigma = \epsilon = 0$ for $t' < 0$,

$$\sigma(t) = \int_0^t E_r(t - t')\frac{d\epsilon(t')}{dt'}\,dt'$$

But, for an engineering stress–strain test

$$\frac{d\epsilon(t')}{dt'} = \dot\epsilon \cong \text{constant} = \dot\epsilon_0$$

Thus,

$$\sigma(t) = E\dot\epsilon_0 \int_0^t e^{-(t-t')/\lambda}\,dt'$$

The integration is performed in the present time (i.e., t is constant) over the material's past history, from $t' = 0$ to $t' = t$, with the result

$$\sigma(t) = E\dot\epsilon_0 \lambda[1 - e^{-t/\lambda}] = \eta_e \dot\epsilon_0 [1 - e^{-(E/\eta_e)t}].$$

This was obtained by direct integration of the differential equation for the Maxwell element in Example 1.

18.7 DYNAMIC MECHANICAL TESTING

Creep and stress-relaxation measurements correspond to the use of step-response techniques to analyze the dynamics of electrical and process systems. Those familiar with these areas know that frequency-response analysis is perhaps a more versatile tool for investigating system dynamics. An analogous procedure, *dynamic mechanical testing*, is applied to the mechanical behaviour of visco-elastic materials. It is based on the fundamentally different response of viscous and elastic elements to a sinusoidally varying stress or strain.

If a sinusoidal strain $\gamma = \gamma'\sin\omega t$ (where ω is the angular frequency, radians per second) is applied to a linear spring, since $\tau = G\gamma$, the resulting stress, $\tau = G\gamma'\sin\omega t$, is *in phase with the strain.* For a linear dashpot, however, because the stress is proportional to the *rate of strain* rather than the strain, $\tau = \eta\dot\gamma = \eta\omega\gamma'\cos\omega t$, the *stress is $90°$ out of phase with the strain.* These relations are sketched in Fig. 18.14.

As might be expected, viscoelastic materials exhibit some sort of intermediate response, which might look like Fig. 18.15b. This can be thought of as being a projection of two vectors, τ^* and γ^*, rotating in the complex plane (Fig. 18.15a). The angle between these vectors is the *phase angle* δ ($\delta = 0$ for a purely elastic material and $90°$ for a purely viscous material). It is customary to resolve the vector representing the dependent variable into components in phase (designated by ') and $90°$ out of phase (designated by ") with the independent variable. In this example the applied *strain* is the independent variable; so the stress vector

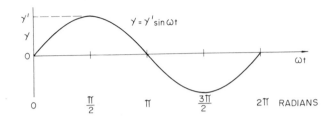

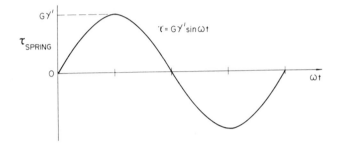

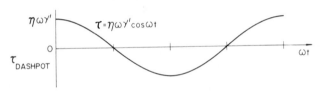

Figure 18.14   Stress in a linear spring and a linear dashpot in response to an applied strain.

(τ^*) is resolved into its in-phase (τ') and out-of-phase (τ'') components, $|\gamma^*| = \gamma'$, and $\gamma'' = 0$.

An *in-phase* or *storage modulus* is defined by

$$G' = \frac{\tau'}{\gamma'} \quad \text{storage modulus} \qquad \text{(in-phase component)} \qquad (18.31)$$

and an *out-of-phase* or *loss modulus* is defined by

$$G'' = \frac{\tau''}{\gamma'} \quad \text{loss modulus} \qquad \text{(out-of-phase component)} \qquad (18.32)$$

The *complex modulus* G^* is defined as the *vector sum* of the in-phase and out-of-phase moduli (using i to denote an out-of-phase unit vector):

$$G^* = G' + iG'' = \frac{\tau' + i\tau''}{\gamma'} = \frac{\tau^*}{\gamma^*} \qquad (18.33)$$

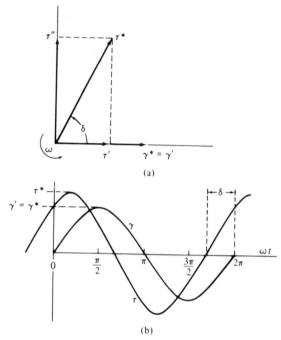

Figure 18.15   Quantities in dynamic testing. (a) rotating vector diagram; (b) stress and strain.

Additionally, a *complex viscosity* η^* may be defined as

$$\eta^* = \eta' - i\eta'' = \frac{\tau^*}{\dot{\gamma}^*} \tag{18.34}$$

It may be shown[8] that

$$\dot{\gamma}^* = i\omega\gamma^* \tag{18.35}$$

By combining (18.33)–(18.35) (and recalling that $i^2 = -1$), we get

$$\eta''\omega + i\omega\eta' = G' + iG'' \tag{18.36}$$

Comparison of the real (in-phase) and imaginary (out-of-phase) parts of (18.36) gives

$$G' = \eta''\omega \tag{18.37}$$

and

$$G'' = \eta'\omega \tag{18.38}$$

Furthermore, combination of (18.34), (18.36), and (18.37) reveals that

$$G^* = G' + iG'' = i\omega\eta^* \tag{18.39}$$

From the geometry of Fig. 18.15 and the relations above, the *loss tangent* $\tan \delta$ is

$$\tan \delta = \frac{\tau''}{\tau'} = \frac{G''}{G'} = \frac{\eta'}{\eta''} \tag{18.40}$$

What is the physical significance of the quantities just defined? This can best be appreciated by considering what happens to the energy applied to a sample undergoing cyclic deformation. The work done on a unit volume of material undergoing a pure shear deformation is

$$w = \int \tau \, d\gamma \qquad (\text{ergs/cm}^3) \tag{18.41}$$

From Fig. 18.15,

$$\gamma = \gamma' \sin \omega t \tag{18.42}$$

and

$$\tau = |\tau^*| \sin (\omega t + \delta) \tag{18.43}$$

Differentiating (18.42) with respect to (ωt)

$$d\gamma = \gamma' \cos \omega t \, d(\omega t) \tag{18.44}$$

Inserting (18.43) and (18.44) into (18.41),

$$w = |\tau^*|\gamma' \int \sin (\omega t + \delta) \cos \omega t \, d(\omega t) \tag{18.45}$$

Let's first consider the work done on the first quarter-cycle of applied strain by integrating (18.45) between 0 and $\pi/2$. Using appropriate trigonometric identities and a good set of integral tables gives

$$w \text{ (first } \tfrac{1}{4} \text{ cycle)} = |\tau^*|\gamma' \left(\frac{\cos \delta}{2} + \frac{\pi}{4} \sin \delta \right) \tag{18.46}$$

In terms of moduli or viscosities, using the trigonometry of Fig. 18.15, (18.37) and (18.38),

$$w \text{ (first } \tfrac{1}{4} \text{ cycle)} = \frac{(\gamma')^2}{2} G' + \frac{\pi}{4} (\gamma')^2 G'' \tag{18.47a}$$

$$w \text{ (first } \tfrac{1}{4} \text{ cycle)} = \frac{(\gamma')^2}{2} \omega \eta'' + \frac{\pi}{4} (\gamma')^2 \omega \eta' \tag{18.47b}$$

The first term on the right-hand side of (18.47a) is simply the work done in straining a linear spring of modulus G' an amount γ'. It therefore represents the energy stored elastically in the material during its straining in the first quarter-cycle. Hence G' is the storage modulus. If the applied mechanical energy (work) is not stored elastically, it must be "lost" — converted to heat through molecular friction, that is, viscous dissipation, within the material. This is precisely what the second term on the right represents; so G'' is known as the loss modulus.

Similarly, from (18.47b), stored energy is proportional to η'' and dissipated energy is proportional to η'.

Considering the second quarter of the cycle, integrating (18.45) from $\pi/2$ to π gives results identical to (18.47), *except that the sign on first (storage) term is negative.* This simply means that the energy stored elastically in straining the material from 0 to γ' is recovered when it returns from γ' to 0. Thus, over a half-cycle or full cycle, there is no net work done or energy lost through the elastic component. The sign of the second term, however, is positive for any quarter-cycle; so the net energy loss (converted to heat within the material) for a full cycle (also obtainable by integrating (18.45) between 0 and 2π) is simply

$$w \text{ (complete cycle)} = \pi(\gamma')^2 G'' = \pi(\gamma')^2 \omega\eta' \qquad (18.48)$$

The average power dissipated as heat within the material, $\langle\dot{w}\rangle$, is obtained by dividing the energy dissipated per cycle by the period (time) of a cycle, $2\pi/\omega$:

$$\langle\dot{w}\rangle \text{ (average power dissipation)} = \tfrac{1}{2}(\gamma')^2 \omega G'' = \tfrac{1}{2}(\gamma')^2 \omega^2\eta' \quad (18.49)$$

These results are of direct importance in the design of polymeric objects that are subjected to cyclic deformation. In a tire, for example, high temperatures contribute to rapid degradation and wear. It is therefore desirable to choose a rubber compound with as low a G'' (or η') as possible to minimize energy dissipation and the resultant heat buildup. In the design of an engine mount, however, the object is usually to prevent the transmission of vibration from the engine. Here, a material with a large G'' (or η') would dissipate much of the vibrational energy as heat rather than transmit it.

Example 7.   Obtain the expressions for the quantities G', G'', $|G^*|$, $\tan\delta$, η', η'' and $|\eta^*|$ for a Maxwell element.

Solution.   This problem is solved in reference 8, p. 56ff by direct integration of the differential equation for the Maxwell element. Here we apply Boltzmann's superposition principle to obtain the results and in so doing illustrate again how one type of linear test (stress relaxation) may be used to predict the response in another (dynamic testing). The shear stress-relaxation modulus (memory function) for a Maxwell element is

$$G(t) = Ge^{-t/\lambda}$$

Therefore

$$G(t-t') = Ge^{-(t-t')/\lambda}$$

We assume that the element has always been subjected to a shear strain

$$\gamma(t') = \gamma'\sin(\omega t')$$

which gives, when differentiated with respect to past time t',

$$\dot{\gamma}(t') = \gamma'\omega\cos(\omega t')$$

(Keep in mind that the prime here has two entirely different meanings. Applied to t, it designates past time; applied to γ, it means the in-phase component of strain.) Plugging these results into (18.30), we get

$$\tau(t) = G\gamma'\omega \int_{-\infty}^{t} e^{-(t-t')/\lambda} \cos(\omega t')\, dt'$$

Evaluating the above integral is nontrivial, but again, with appropriate trigonometric identities and a good table of integrals, the result is

$$\tau(t) = \frac{G\gamma'(\omega\lambda)^2}{1+(\omega\lambda)^2} \sin(\omega t) + \frac{G\gamma'\omega\lambda}{1+(\omega\lambda)^2} \cos(\omega t)$$

(By assuming that the element has *always* been subjected to the periodic strain, i.e., integrating from $t' = -\infty$, we eliminate the transient part of the solution that would arise had we integrated from $t' = 0$.) It is clear that the first (sin) term in the above expression is *in phase* with the applied strain, while the second (cos) term is 90° out of phase with the applied strain. Therefore

$$\tau' = \frac{G\gamma'(\omega\lambda)^2}{1+(\omega\lambda)^2} \qquad \text{and} \qquad \tau'' = \frac{G\gamma'\omega\lambda}{1+(\omega\lambda)^2}$$

From here on, it's definitions and algebra:

$$G' = \frac{\tau'}{\gamma'} = \frac{G(\omega\lambda)^2}{1+(\omega\lambda)^2} \qquad \text{and} \qquad G'' = \frac{\tau''}{\gamma'} = \frac{G\omega\lambda}{1+(\omega\lambda)^2}$$

$$|G^*| = [(G')^2 + (G'')^2]^{1/2} = \frac{G\omega\lambda}{[1+(\omega\lambda)^2]^{1/2}}$$

$$\tan\delta = \frac{G''}{G} = \frac{1}{\omega\lambda}$$

$$\eta' = \frac{G''}{\omega} = \frac{G\lambda}{1+(\omega\lambda)^2} = \frac{\eta}{1+(\omega\lambda)^2}$$

$$\eta'' = \frac{G'}{\omega} = \frac{G\lambda^2}{1+(\omega\lambda)^2} = \frac{\eta\omega\lambda}{1+(\omega\lambda)^2}$$

$$|\eta^*| = [(\eta')^2 + (\eta'')^2]^{1/2} = \frac{\eta}{[1+(\omega\lambda)^2]^{1/2}}$$

The dynamic moduli are plotted in dimensionless form in Fig. 18.16a, and the dynamic viscosities in Fig. 18.16b.

As when applied to other types of mechanical tests, the Maxwell element won't win any prizes for quantitatively fitting dynamic data for real materials.

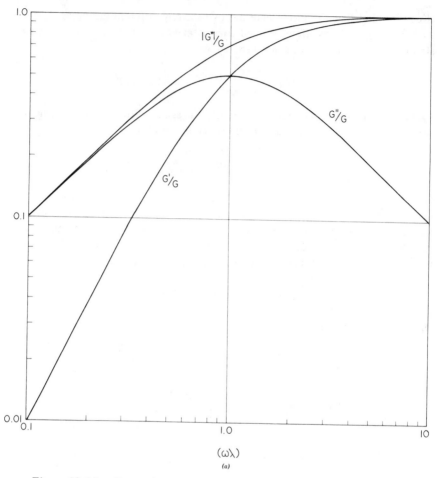

Figure 18.16   Dynamic properties of a Maxwell element (Example 6).

Nevertheless, Example 7 does serve to illustrate the frequency dependence of dynamic mechanical properties, and Figs. 18.16a and 18.16b do in some ways resemble the variation in isothermal dynamic data with frequency for real materials. In particular, the apparent "stiffness" $|G^*|$ increases to a limiting value with frequency. In the model the dashpot simply can't keep up with high frequencies, leaving only the spring to respond. Also, the maximum in G'' is usually observed at frequencies in the range where G' and $|G^*|$ are falling from their high-frequency limit. In the model, at low frequencies, the dashpot offers little resistance to motion, and so dissipates little energy, while at high frequencies, its high resistance prevents its motion, and energy dissipation again falls off.

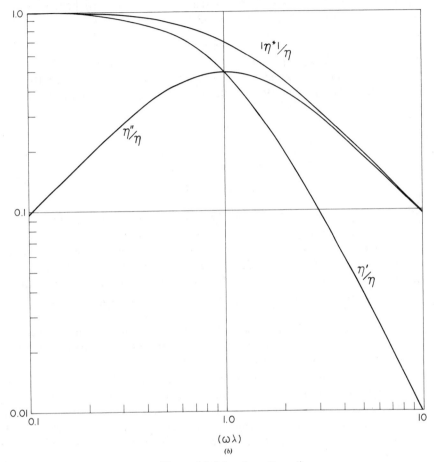

Figure 18.16   (continued)

Experimentally, it is observed that the high-frequency limit of $|G^*|$ and G' corresponds quantitatively to the moduli obtained in the limit of short times in creep and stress-relaxation measurements; that is,

$$|G^*|(\omega \to \infty) = G'(\omega \to \infty) = G(t \to 0) = \frac{1}{J(t \to 0)} \qquad (18.50)$$

Furthermore, the measured low-frequency limit of $|\eta^*|$ and η' agrees with the zero-shear rate steady-flow viscosity η_0

$$|\eta^*|(\omega \to 0) = \eta'(\omega \to 0) = \eta_0 = \eta(\dot{\gamma} \to 0) \qquad (18.51)$$

The viscosity analogy can be pushed a bit further. The drop in $|\eta^*|$ with frequency resembles the variation in steady-flow viscosity η, with shear rate.

On a purely empirical basis, Cox and Merz[9] suggested that $|\eta^*|$ and η are the same when compared at equal values of frequency and shear rate:

$$|\eta^*|(\omega) = \eta(\dot{\gamma}) \quad at \quad \omega = \dot{\gamma} \tag{18.52}$$

While not exact, (18.52) appears to be at least a reasonable approximation.

For a generalized Maxwell model consisting of n elements (Fig. 18.10), the results obtained for a single Maxwell element in Example 7 are readily generalized to

$$G' = \sum_{i=1}^{n} \frac{G_i(\omega\lambda_i)^2}{1 + (\omega\lambda_i)^2} \tag{18.53}$$

and

$$G'' = \sum_{i=1}^{n} \frac{G_i(\omega\lambda_i)}{1 + (\omega\lambda_i)^2} \tag{18.54}$$

(the other dynamic properties may be obtained from the two above, as in Example 7). Presumably, the G_i and λ_i determined from stress-relaxation data, for example, can be used in (18.53) and (18.54) to predict dynamic response, and vice versa. The relations (18.27) and (18.28) should be equally applicable to both. Also, the continuous distributions $G(\lambda)$ and $J(\lambda)$ as obtained from stress-relaxation and creep measurements are at least approximately inter-convertible with $G'(\omega)$ and $G''(\omega)$.[2,4]

Basically, three methods are available for determining dynamic properties: free oscillation, forced oscillation, and steady-state rotation. Experimental and analytical details for the first two are reviewed extensively by Ferry.[2]

Free-oscillation measurements are made with a torsion pendulum (Fig. 18.17). The sample is given an initial torsional displacement, and the frequency and amplitude decay of the oscillations are observed on release. G' is determined from the sample geometry, moment of inertia of the oscillating mechanism, and the observed period of oscillation. For example, with a cylindrical specimen of length L and radius R,

$$G' = \frac{8\pi L I}{R^4 P^2} \tag{18.55}$$

where I is the moment of inertia of the oscillating mechanism and P is the observed period of oscillation ($P = 2\pi/\omega$). The damping, or logarithmic decrement Δ is calculated from the amplitude decay of the oscillations:

$$\Delta = \ln\frac{A_1}{A_2} = \ln\frac{A_2}{A_3} = \ln\frac{A_i}{A_{i+1}} = \frac{1}{n}\ln\frac{A_i}{A_{i+n}} \tag{18.56}$$

and for $\Delta < 1$

$$\Delta \cong \pi \tan \delta \tag{18.57}$$

Expressions for other specimen geometries and higher damping are reviewed by Nielsen.[5] Solid or rubbery samples are twisted, as illustrated in Fig. 18.17, in

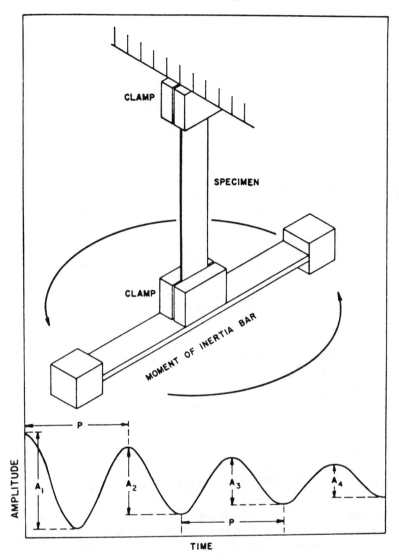

Figure 18.17   Torsion pendulum with its output.[17]

the form of rods, tubes, strips, etc. Liquids or soft solids may be contained in one of the geometries described for rotational viscometry in Chapter 16 (Couette, cone-and-plate, etc.). In *torsional braid analysis*,[10] a flexible braided fiber – usually glass – is impregnated with the material to be studied. This impregnated fiber then becomes the torsion member in the pendulum. In analyzing the data from such a device, care must be exercised in separating interactions between the sample and supporting fiber.

Although the frequency can be varied somewhat by changing the moment of inertia of the oscillating portion of the mechanism, torsion pendula are usually intended to study only the temperature dependence of dynamic properties at a constant, relatively low frequency. On the other hand, they are inexpensive and rather simple to construct.

Forced-oscillation devices ("jiggle machines") apply a sinusoidal stress or strain of known amplitude and frequency and measure the resulting strain or stress. The dynamic properties are calculated from the relation between the two. For liquid samples, the geometries discussed in conjunction with rotational viscometry in Chapter 16 are often used with the drive system modified to produce sinusoidal rather than steady rotational deformation. Flexible samples such as fibers, films, and rubber are preloaded in tension and oscillated about a positive tensile strain so that they don't go slack at the "bottom" of the sine wave. Such tests give dynamic *tensile* properties E', E'', and so on, which are related to the corresponding shear properties by

$$|E^*| = 2|G^*|(1 + v) \tag{18.58}$$

where v is Poisson's ratio ($v = 1/2$ for an incompressible material). Another type of forced-oscillation device applies a sinusoidal shear or compression wave to one end of a sample and monitors the attenuation of the wave as it progresses through the sample.

Forced-oscillation devices are generally intended to study dynamic properties as functions of frequency as well as of temperature. Drive and detection systems for such devices may be strictly mechanical, but the newer and more sophisticated systems make use of piezoelectric crystals, inorganic crystals (e.g., barium titanate) that change dimension in proportion to an applied voltage and, conversely, generate an output voltage proportional to an imposed deformation. Thus, the amplitude and especially the frequency of the applied strain can be conveniently controlled over a wide range in the form of electrical signals (the response of the crystals extends to very high frequencies, and electrons have very little inertia, unlike mechanical linkages), and input and output may be compared directly on an oscilloscope.

The expense and complexity of oscillating drives and associated detection systems have led to the increased popularity of instruments that determine dynamic properties but are driven in steady rotation and apply or detect steady (*non*oscillating) forces or deformations. This sounds like a contradiction in terms, but they really do work! Unfortunately, these devices are based on nonviscometric deformations, that is, two-dimensional variation in stress and strain, which complicates analysis. Nevertheless, we shall attempt to describe the

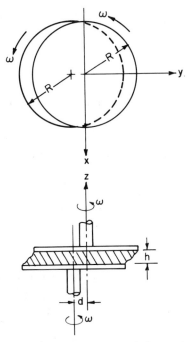

Figure 18.18   Orthogonal rheometer.

operation of one of the more popular such devices. Mathematical details are provided in elegant fashion by Walters.[11]

In the Maxwell *orthogonal rheometer*[12] (Fig. 18.18), material is sheared between two parallel disks of radius R, separated from one another by a distance h, each rotating at the same angular frequency ω, *but with their axes of rotation displaced* by a distance d. Transducers are set up to measure three steady orthogonal forces (hence the name) on one of the disks, F_x, F_y, and F_z.

How, you ask, does this result in oscillating deformation? Well, the easiest way to see this is to follow the motion of a point on the upper disk (point 2) relative to a point (point 1) on the lower disk as the disks rotate. Here, we choose point 1 on the axis of rotation of the lower disk and point 2 directly above it at the start of the analysis, $\omega t = 0$. This is the simplest choice, because point 1 remains stationary, but the same result will be obtained for any pair of points initially at the same values of x and y (you can prove this with a compass, ruler, and protractor). Figure 18.19 shows the relative displacement vector l in the x–y plane, and how it varies with the angle of rotation ωt. From Fig. 18.19b, the magnitude of the relative displacement is then

$$|\mathbf{l}| = 2d \sin \frac{\omega t}{2} \qquad (18.59)$$

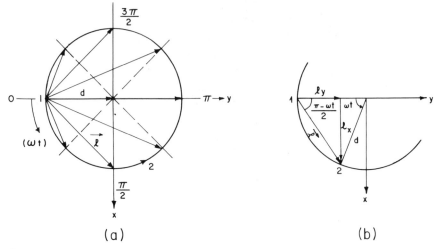

(a)   (b)

Figure 18.19   Analysis of the orthogonal rheometer. (*a*) relative displacement vector between point 1 on center of lower disk and point 2 immediately above it at $\omega t = 0$; (*b*) geometry of the displacement vector and its components.

and its x and y components are

and
$$l_x = d \sin (\omega t) \tag{18.60}$$

$$l_y = |l| \cos \left(\frac{\pi}{2} - \frac{\omega t}{2} \right) = d[1 - \cos (\omega t)] \tag{18.61}$$

If we assume that the strain is simply the relative displacement divided by the disk separation h, the x and y components of strain are*

$$\gamma_x = \frac{d}{h} \sin (\omega t) \tag{18.62}$$

$$\gamma_y = \frac{d}{h}[1 - \cos (\omega t)] \tag{18.63}$$

Thus both components of strain undergo simple harmonic oscillation as a result of the steady angular rotation.

It is important to note that the x component of strain is symmetrical, while the y component is not. For linear materials, because shear stress is linear in strain, any shear stresses, and therefore forces parallel to the x–y plane produced

*This requires that each fluid element move at fixed z in a circle that is centered on a line connecting the centers of the disks. There has been some controversy about this, but it is probably at least a good approximation for $d/h < 0.5$.

by γ_x's in the region $(0 < \omega t < \pi)$, will be canceled by those arising from the equal-and-opposite x-component strains in the region $(\pi < \omega t < 2\pi)$. As a result, the measured forces F_x and F_y can depend only on γ_y. Looking at it another way, consider a purely elastic material in the rheometer represented by rubber bands stretched between points 1 and 2. Viewed from above, the rubber bands coincide with 1. It is obvious that the x components of the tug of the rubber bands will cancel, leaving only a net force (and therefore only a stress component τ_{zy}) in the y direction. Thus the purely elastic stress is in phase with γ_y, and (18.63) represents the variation of γ' with location in a disk at constant z; that is,

$$\gamma'(\omega t) = \gamma_y(\omega t) = \frac{d}{h}[1 - \cos(\omega t)] \tag{18.64}$$

and

$$\tau_{zy} = \tau' = \tau'(\gamma') \tag{18.65}$$

Any measured value of F_x, therefore, must arise from a viscous component, whose stress τ_{zx} will be $90°$ out of phase with γ_y; that is,

$$\tau_{zx} = \tau''(\gamma'). \tag{18.66}$$

In order to determine the total force component on a disk, the stress–area product must be integrated over the surface of the disk:

$$F_y = \int_{\substack{\text{surface} \\ \text{of disk}}} \tau' \, dA \tag{18.67}$$

From (18.31), $\tau' = G'\gamma'$, and in polar coordinates, $dA = r \, d(\omega t) \, dr$, so that

$$F_y = G'\left(\frac{d}{h}\right) \int_0^R \int_0^{2\pi} [1 - \cos(\omega t)] \, d(\omega t) r \, dr = \pi R^2 \left(\frac{d}{h}\right) G' \tag{18.68}$$

or

$$G' = \frac{(h/d)}{\pi R^2} F_y \tag{18.69}$$

Similarly,

$$F_x = \int_{\substack{\text{surface} \\ \text{of disk}}} \tau'' dA \tag{18.70}$$

and since $\tau'' = G''\gamma'$ (18.32), it follows that

$$G'' = \frac{(h/d)}{\pi R^2} F_x \tag{18.71}$$

The third force, F_z, is believed related to the second normal stress difference σ_2 (17.4b).

Equations (18.69) and (18.71) show how dynamic properties can be obtained from a device in steady rotation by measuring steady forces, which mechanically and analytically represents a great simplification over forced-oscillation techniques. It must be pointed out that there have been some questions as to whether a nonviscometric flow such as this can be used to determine quantities such as G', G'', η', η'', and σ_2, which are really defined in terms of viscometric deformations, but it is now pretty generally agreed that the technique is valid, at least in the limit of small strains d/h.[13,14] Also, analyses are based on the assumption that both disks rotate at the same ω. In the usual instrument, one disk is driven and the other goes along for the ride. With bearing friction and hydrodynamic effects, this assumption might not always be true. An analysis by Davis and Macosko shows it to be pretty good under most conditions of interest, however.[15]

Other devices that operate along similar lines have been developed. The so-called balance rheometer confines a test fluid between two concentric hemispheres that rotate at the same rate but whose axes of rotation are inclined at an angle to one another. Similarly, cone-and-plate geometry, in which the axis of the cone is not perpendicular to the plate, and Couette (cup-and-bob) geometry, with the bob not centered in the cup, can also be used to obtain dynamic data. These devices are analyzed by Walters.[11] Dynamic tensile properties can be obtained for relatively rigid materials by subjecting a rotating cylindrical rod to a cantilever deflection.[16]

Figure 18.20 illustrates G' and damping $\cong \pi \tan \delta$ versus T for polymethyl methacrylate, an amorphous linear polymer. The data were obtained with a torsion pendulum at about 1 cycle/sec. At low temperatures, the typical glassy modulus, 10^{10} to 10^{11} dynes/cm^2 ($10^9 - 10^{10}$ N/m^2), is observed. In the vicinity of $110°$ to $130°$C, G' drops precipitously, ultimately reaching a plateau of 10^6 to 10^7 dynes/cm^2 ($10^5 - 10^6$ N/m^2), the typical rubbery modulus. Also, a sharp peak in the damping is observed in this region. Although the temperature at which this peak and drop are observed is frequency dependent, at low frequencies (such as are obtained with the usual torsion pendulum), they are identified with the material's glass-transition temperature, and the drop in G' is indicative of the decrease in "stiffness" at T_g, going from the straining of bond angles and lengths to coiling and uncoiling as the dominant response mechanism. The damping peak represents the onset of cooperative motion of 40 to 50 main-chain carbon atoms at T_g (Chapter 8). In terms of the four-parameter model, below T_g only spring 1 is operative, and the material is almost completely elastic (low damping). In the vicinity of T_g the viscosity of dashpot 2 drops to the point where it can deform and dissipate energy, giving the damping peak. At higher temperatures its viscosity drops to the point where it dissipates little energy, and the material is again

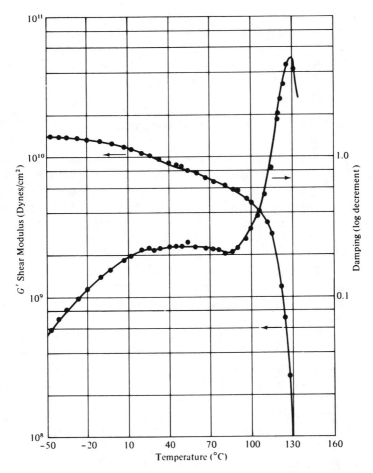

Figure 18.20   Dynamic mechanical properties of polymethyl methacrylate.[17] The data were obtained with a torsion pendulum at about 1 cycle/sec.

highly elastic, mainly through spring 2. At still higher temperatures the modulus drops off rapidly owing to viscous flow (dashpot 1). The broad damping "hill" centered at about 40°C (and the accompanying gradual drop in G') in Fig. 18.20 has been shown to arise from the motion of the $-\overset{\overset{\displaystyle O}{\parallel}}{C}-O-CH_3$ side groups.

Note that the dimensions of the angular frequency ω are sec^{-1}. Thus ω corresponds to a *reciprocal* time scale. For dynamic (oscillating) deformations, then, the Deborah number is

$$De = \lambda_c \omega \qquad (18.72)$$

In dynamic tests, a viscoelastic material will become more solidlike as the frequency is increased and the time-dependent response mechanisms become unable to follow the rapidly reversing stress. In the limit of very high frequencies, the straining of bond angles and lengths will be the only operative response mechanism, and the polymer will exhibit the typical glassy modulus even though it may be well above its glass-transition temperature.

18.8   TIME–TEMPERATURE SUPERPOSITION

Anyone who has ever wrestled with a cheap garden hose in cold weather appreciates the fact that polymers become stiffer and more rigid at lower temperatures, while at high temperatures they are softer and more flexible. In preceding examples, we have seen that the time scale (or frequency) of the application of stress has a similar influence on mechanical properties, short times (or high frequencies) corresponding to low temperatures and long times (low frequencies) to high temperatures. A ball of "silly putty," if heated sufficiently above room temperature (assuming it did not degrade first), would simply stay on the floor like clay when dropped. Conversely, if cooled sufficiently below room temperature and stuck on the wall, it would exhibit mainly elastic deformation for long periods of time.

The quantitative application of this idea, *time–temperature superposition*, is one of the most important principles of polymer physics. It is based on the fact that the Deborah number determines quantitatively just how a viscoelastic material will behave mechanically. Changing *either* t_s (or ω) *or* λ_c can change De. The nature of the applied deformation determines t_s (or ω), *while a polymer's characteristic time is a function of temperature.* The higher the temperature, the more thermal energy the chain segments possess, and the more rapidly they are able to respond, lowering λ_c. Thus, for example, De can be doubled by halving t_s (or doubling ω in a dynamic test) *or* by lowering the temperature enough to double λ_c. The change in mechanical response will be the same either way, according to the time–temperature superposition principle.

Although time–temperature superposition is applicable to the widest variety of viscoelastic response tests (creep, dynamic, etc.) its application is illustrated here with stress relaxation. Figure 18.21 shows tensile stress relaxation data at various temperatures for polyisobutylene, plotted in the form of a time-dependent tensile (Young's) modulus $E_r(t)$ versus time on a log–log scale.

$$E_r(t) = \frac{f(t)/A}{\Delta l/l \text{ (constant)}} \qquad (18.73)$$

where $f(t)$ = the measured tensile force in the sample.

    A = its cross-sectional area.

and l = its length.

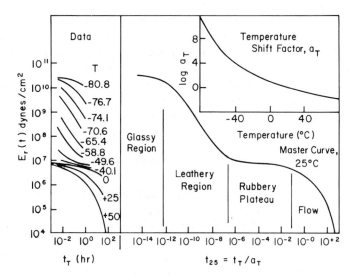

Figure 18.21 Time–temperature superposition for NBS polyisobutylene. Adapted from reference 18.

The technique is, of course, equally applicable to shear deformation. In stress relaxation, the lower measurement time limit is set by the assumption that the constant strain is applied instantaneously. In practice, inertia and other mechanical limitations make this impossible; so data are only valid at times a couple of order of magnitude longer than it actually takes to apply the constant strain. The upper limit is set by the dedication of the experimenter and the long-term stability of the sample and equipment. These data were obtained over a range of seconds to a couple of days. As might be expected, the modulus drops with time at a given temperature, and at a given time it drops with increasing temperature.

If we stare at the curves for a while, they appear to be sections of one continuous curve, chopped up, with the sections displaced along the log–time axis. That this is indeed so is shown in Fig. 18.21. Here, 25°C has *arbitrarily* been chosen as a reference temperature T_0 and the curves for other temperatures shifted along the log–time axis to line up with it. The data below 25° must be shifted to the left (shorter times) and those above 25°, to the right (longer times), giving a *master curve* at 25°.

Sometimes the relaxation moduli at each temperature T are corrected to the reference temperature T_0 by multiplying by the ratio T_0/T before superposing. This correction is based on the theory of ideal rubber elasticity (Chapter 14), which states that the modulus is proportional to the absolute temperature. This procedure is open to question, however. While ideal rubber elasticity might be a reasonable approximation when the major response mechanism is chain coiling and uncoiling, it certainly is not where response is dominated by straining of bond angles and lengths (glassy region) or molecular slippage (viscous flow. region). In any event, such corrections are of minor practical significance.

Shifting a constant-temperature curve along the log–time axis corresponds to dividing every value of its abscissa by a constant factor (it is immaterial what kind of scale is used for the ordinate). This constant factor, which brings a curve at a particular temperature into alignment with the one at the reference temperature, is known as the *temperature shift factor* a_T:

$$a_T = \frac{t_T}{t_{T_0}} \qquad \text{(for same response)} \qquad (18.74)$$

where t_T is the time required to reach a particular response (modulus, in this case) at temperature T and t_{T_0} is the time required to reach the *same response* at the reference temperature T_0. For temperatures above the reference temperature, it takes less time to reach a particular response (the material responds faster, i.e., has a shorter relaxation time), so a_T is less than 1, and vice versa. The logarithm of the experimentally determined temperature shift factor is plotted as a function of temperature in Fig. 18.21.

The master curve now represents stress relaxation at 25° *over 17 decades of time*. Multiplication by the appropriate value of a_T establishes the master curve at any other temperature and can thus be used to predict response at that temperature over the entire time scale.

There are two additional aspects that enhance the utility of the time–temperature superposition concept. First, the same temperature shift factors apply to a particular polymer regardless of the nature of the mechanical response; that is, the shift factors as determined in stress relaxation are applicable to the prediction of the time–temperature behaviour in creep or dynamic testing. Second, *if the polymer's glass-transition temperature is chosen as the reference temperature*, the shift factors are given, to a good approximation, by the Williams–Landel–Ferry (WLF) equation in the range $T_g < T < (T_g + 100°C)$:

$$\log a_T = \frac{-17.44(T - T_g)}{51.6 + (T - T_g)} \qquad \text{(for } T_0 = T_g) \qquad (18.75)$$

Example 8. The damping for polymethyl methacrylate in Fig. 18.20 is located at 130°C. Assuming that the data were obtained at a frequency of 1 cycle/sec, at what temperature would the peak be located if measurements were made at 1000 cycles/sec? For polymethyl methacrylate, $T_g = 105°C$.

Solution. Keeping in mind that frequency is a *reciprocal* time scale, and applying the WLF equation,

$$\log a_T = \log \frac{\omega_{T_g}}{\omega_T} = \frac{-17.44(T - T_g)}{51.6 + T - T_g}$$

Shifting the measurements at 1 cycle/sec to T_g (i.e., finding the frequency at which the peak would be located at T_g),

$$\log \frac{\omega_{105°}}{\omega_{130°}} = \frac{-17.44(130 - 105)}{51.6 + 130 - 105} = -5.69$$

$$\frac{\omega_{105°}}{\omega_{130°}} = 2.03 \times 10^{-6}$$

and

$$\omega_{105°} = (1 \text{ cycle/sec})(2.03 \times 10^{-6}) = 2.03 \times 10^{-6} \quad \text{cycles/sec}$$

Now, shifting from T_g to T,

$$\log \frac{2.03 \times 10^{-6}}{10^3} = -8.69 = \frac{-17.44(T - 105)}{51.6 + T - 105}$$

$$T = 156°C$$

This illustrates quantitatively the statements made at the end of Section 18.7.

Example 9.   The master curve for the polyisobutylene in Fig. 18.21 indicates that stress relaxes to a modulus of 10^6 dynes/cm^2 in about 10 hr *at 25°C.* Using WLF equation, estimate the time it will take to reach the same modulus at a temperature of $-20°C$. For PIB, $T_g = -70°C$.

Solution.   To use the WLF equation, the reference temperature must be $T_g = -70°C$.

$$\log \left(\frac{t_T}{t_{T_g}} \right) = \log \frac{t_{25°}}{t_{-70°}} = \frac{-17.44(25 + 70)}{51.6 + 25 + 70} = -11.3$$

$$\frac{t_{25°}}{t_{-70°}} = 5.01 \times 10^{-12} \quad \text{and} \quad t_{-70°} = \frac{10}{5.01 \times 10^{-12}} = 2 \times 10^{12} \quad \text{hr}$$

$$\log \frac{t_{-20°}}{t_{-70°}} = \frac{-17.44(-20 + 70)}{51.6 - 20 + 70} = -8.59$$

$$\frac{t_{-20°}}{t_{-70°}} = 2.57 \times 10^{-9}$$

$$t_{-20°} = (2 \times 10^{12})(2.57 \times 10^{-9}) = 5140 \quad \text{hr}$$

This shows how lowering the temperature maintains mechanical "stiffness" for much longer periods of time.

Let's take a closer look at the stress-relaxation master curve. At low temper-

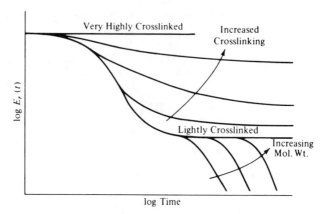

Figure 18.22 The effects of molecular weight and crosslinking on stress-relaxation master curves.

atures or short times, only bond angles and lengths can respond, and so the typical glassy modulus, 10^{10} to 10^{11} dynes/cm^2 ($10^9 - 10^{10}$ N/m^2), is observed. This is the so-called *glassy region*. At longer times or higher temperatures, response is governed by the uncoiling of the chains, with the characteristic 10^6 to 10^7 dynes/cm^2 ($10^5 - 10^6$ N/m^2) in the *rubbery plateau*. The intermediate region, where the modulus drops from glassy to rubbery, is sometimes known as the *leathery region* from the leatherlike feel of polymers in this region. At still longer times, the modulus drops off rapidly as a result of molecular slippage — the *viscous-flow region*. Figure 18.22 illustrates the effect of molecular weight and crosslinking on stress relaxation. Molecular weight should have no influence on straining of bond angles and lengths or on uncoiling; so the glassy and rubbery moduli are unchanged. Viscous flow, however, is severely retarded by increasing molecular weight; in the limit of infinite molecular weight (light crosslinking), it is eliminated entirely, and the curve levels off with the rubbery modulus. Higher degrees of crosslinking restrict uncoiling, ultimately leading to a material that responds only by straining of bond angles and lengths.

The effects of crystallinity on properties are similar to those of crosslinking. However, the applicability of time–temperature superposition in and across the region of T_m is questionable, at best. In this region, the degree of crystallinity and crystalline morphology change, and one would be, in effect, superposing data for different materials. A second, vertical shift has been suggested to help superpose data for crystalline polymers.

To illustrate the effect of temperature on mechanical properties, it is sometimes preferable to plot the property versus temperature for constant values of time. For example, the data in Fig. 18.22 are sometimes cross-plotted as $E_r(T)$ versus T at $t = 10$ sec (the 10-second modulus).[4]

REFERENCES

1. M. Reiner, *Phys. Today*, Jan. 1964, p. 62.
2. J. D. Ferry, *Viscoelastic Properties of Polymers*, 3rd ed., Wiley, New York, 1980.
3. F. R. Eirich, Ed., *Rheology*, Vols. 1–4, Academic Press, New York, 1956–1964.
4. A. V. Tobolsky, *Properties and Structure of Polymers*, Wiley, New York, 1960.
5. L. E. Nielsen, *Mechanical Properties of Polymers and Composites*, Vol. 1, Marcel Dekker, New York, 1974.
6. T. W. Spriggs, *Chem. Eng. Sci.*, **20**, 931 (1965).
7. P. E. Rouse, Jr., *J. Chem. Phys.*, **24**, 269 (1956).
8. J. M. McKelvey, *Polymer Processing*, Wiley, New York, 1962.
9. W. P. Cox and E. H. Merz, *J. Polym. Sci.*, **28**, 619 (1958).
10. J. K. Gillham, *Am. Inst. Chem. Eng. J.*, **20**, 1066 (1974).
11. K. Walters, *Rheometry*, Halsted Press, New York, 1975.
12. B. Maxwell and R. P. Chartoff, *Trans. Soc. Rheol.* **9**, 41 (1965).
13. R. B. Bird and E. K. Harris, *Am. Inst. Chem. Eng. J.*, **14**, 758 (1968).
14. C. W. Macosko, and W. M. Davis, *Rheol. Acta*, **13**, 814 (1974).
15. W. M. Davis and C. W. Macosko, *Am. Inst. Chem. Eng. J.*, **20**, 600 (1974).
16. B. Maxwell, *J. Polym. Sci.*, **20**, 551 (1956).
17. L. E. Nielsen, *SPE J.*, **16**, 525 (1960).
18. A. V. Tobolsky and E. Catsiff, *J. Polym. Sci.*, **19**, 111 (1956).

PART 4

Polymer Technology

CHAPTER 19

Processing

19.1 INTRODUCTION

The term *processing* is used here to describe the technology of converting raw polymer, or compounds containing raw polymer, to articles of a desired shape. Considering the wide variety of polymer types and the even wider variety of articles made from them, a complete description and analysis of the myriad of processing techniques that have sprung up through the years would be impossible here. There are several references that devote considerable space to detailed descriptions of polymer processing operations.[1-4] Similarly, quantitative treatment of processing operations has been the subject of a number of books.[5-12] Here, we outline the common processing techniques, introduce relevant terminology, and consider how the various techniques are based on the fundamentals previously discussed. The reader wishing additional detail may consult the references.

19.2 MOLDING

Molding consists of confining a material in the fluid state to a mold where it solidifies, taking the shape of the mold cavity.

Injection molding is the most common means of fabricating thermoplastic articles. Figure 19.1 illustrates a contemporary injection press. The molding compound, usually in the form of pellets that are approximately 1/8-in. cubes or cylinders 1/8-in. in diameter by 1/8-in. in length (molding "powder"), is fed from a hopper (which may be heated with circulating hot air to dry the material) to an electrically heated barrel. The pellets are conveyed forward through the barrel by a rotating screw. The material is melted as it goes by a combination of heat from the barrel and the shearing (viscous energy dissipation) of the screw.

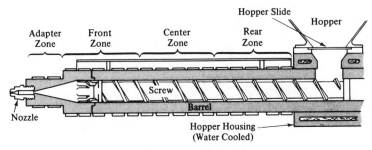

Figure 19.1   Injection Molding Machine (Modern Plastics Encyclopedia, McGraw-Hill, 1969–1970 ed.).

Older machines had a reciprocating plunger in place of the screw. All the heat for melting had to be supplied by conduction from the barrel, and mixing was very poor. This resulted in low plasticizing (melting) rates and a thermally non-uniform melt. The screw generates heat *within* the material and provides some mixing, thereby increasing plasticizing rates and giving a much more uniform melt temperature. Molten material passes through a check valve at the front of the screw, and as it is deposited ahead of the screw, it pushes the screw backward while material from the previous shot is cooling in the mold. The cooled parts are ejected from the mold, the mold closes, and the screw is pushed forward hydraulically, injecting a new shot into the mold.

The molten polymer flows through a *nozzle* into the water-cooled mold, where it travels in turn through a *sprue, runners*, and a narrow *gate* into the *cavity* (Fig. 19.2). When the part has cooled sufficiently, the mold opens, the knockout pins eject the parts, with sprues and runners attached. The sprues and runners are usually removed by hand and, with thermoplastics, chopped and recycled back to the hopper as *regrind*. Depending on the extent of material degradation in the cycle and the property requirements of the finished part, injection molding operations may tolerate up to 25% regrind in the hopper feed.

The molds are opened and closed by hydraulic cylinders, toggle mechanisms, or combinations of both. These have to be pretty hefty because pressures in the mold can reach many thousand pounds per square inch. The molds themselves often involve much intricate hand labor and thus can be quite expensive, but when amortized over a production run of many parts, the contribution of mold cost to the cost of the finished item may be insignificant.

Injection molding presses are rated in terms of tons of mold-clamping capacity and ounces (of general-purpose polystyrene) of shot size. They range from 2-ton, 0.25-oz laboratory units to 3500-ton, 1500-oz monsters that are used to mold garbage cans, TV cabinets, furniture, and so forth. They are usually completely automated. Cycle times (and therefore production rates) sometimes depend on plasticizing capacity (the rate at which the material can be melted), but more often than not they are limited by the cooling time in the mold, which is in

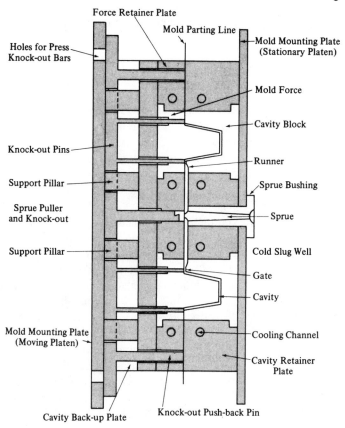

Force Retainer Plate

Mold Parting Line

Holes for Press Knock-out Bars

Mold Mounting Plate (Stationary Platen)

Mold Force

Cavity Block

Knock-out Pins

Runner

Support Pillar

Sprue Bushing

Sprue Puller and Knock-out

Sprue

Support Pillar

Cold Slug Well

Gate

Cavity

Mold Mounting Plate (Moving Platen)

Cooling Channel

Cavity Retainer Plate

Cavity Back-up Plate

Knock-out Push-back Pin

Figure 19.2   Two-cavity Injection Mold (Modern Plastics Encyclopedia, McGraw-Hill, 1969−1970 ed.).

turn established by the thickness of the part and the (usually very low) thermal diffusivity of the material. Typical cycle times are between $\frac{1}{4}$ and 2 min. As usual, there are compromises involved. It is often tempting to try to reduce cooling time by lowering the mold temperature, thereby increasing the cooling rate, and/or by lowering the temperature at which the material is injected into the mold, reducing the amount it must be cooled before solidying enough to allow removal from the mold without distortion. The higher material viscosity that results, however, can give rise to short shots (incomplete mold filling), poor surface finish, and part distortion from frozen-in strains (see below).

Injection molding machines are increasingly being equipped with feedback control systems that monitor cavity pressure, screw position, and/or screw velocity and control these quantities through the hydraulic system to provide high-quality, uniform parts, using a minimum amount of material and minimizing the number of rejects.

Example 1.   Use the four-parameter model (Fig. 18.2f) to explain the presence of frozen-in strains in injection-molded parts.

Solution.   When a polymer melt is squirted into a mold, it is subjected to a combination of volumetric (from the high pressures), shear, and elongational (from flow-induced orientation) strains. On a molecular level, these all arise because of deviations from the most random, highest entropy configuration and therefore can be represented by extension of the Voigt–Kelvin element, spring 2 and dashpot 2 of the four-parameter model. When the mold has filled and the gate freezes, these strains begin to relax (the Voigt–Kelvin element begins to contract). However, before contraction is complete, the rapidly dropping temperature in the mold raises η_2 to the point where the rate of retraction becomes very low, and the part is ejected from the mold with the strain frozen in. Subsequent recovery outside the mold causes distortion of the part. The problems are compounded by the fact that temperatures and flow fields are not uniform within the mold; so the frozen-in strains vary from point to point in the part, giving rise to nonuniform distortions (warping). (It is for this reason that the thermally uniform melt from a screw machine is beneficial.) These problems can, of course, be minimized by using higher melt temperatures and/or mold temperatures, but both lengthen cycle time.

It might also be pointed out here that part failure can result if these frozen-in strains are suddenly released, as, for example, by the concentration of stress by a scratch or sharp corner on the part, or by plasticization by a liquid (stress cracking).

Foamed plastic articles are molded by incorporating a blowing agent (Chapter 20). These blowing agents may be inert but volatile chemicals (e.g., fluorocarbons), compounds that decompose chemically at elevated temperatures to liberate a gas such as nitrogen or carbon dioxide, or simply nitrogen dissolved in the polymer. In any case, the blowing gas is kept in solution by the high pressures ahead of the screw. The mold is partially filled with a shot from the cylinder. The reduced pressure within the mold allows the blowing agent to vaporize, expanding the shot to fill the cavity.

Injection molding formerly was confined exclusively to thermoplastics. It is now also used for thermosets, which are injected into a heated mold where they solidify through the curing (crosslinking) reaction. This is a tricky operation. The compound must be heated just enough in the barrel to achieve fluidity and injected into the mold before it begins to cure appreciably. This requires precise control of temperatures and cycle timing to prevent premature cure in the barrel and the resulting shutdown and cleanout operation. Thermoset sprues and runners cannot be recycled, of course, and must be minimized to cut waste.

Thermosetting compounds are traditionally *compression* molded. The molds are mounted in hydraulic presses on steam-, electric-, or oil-heated *platens*. The molding compound is fed to the heated mold, which closes, maintaining the material under pressure until cured. The part is then ejected from the mold.

Molding compound in the form of granules or powder may be fed to the mold automatically in weighed shots or as preformed (by cold pressing) tablets. The charge is often preheated to reduce heating time in the mold and thereby cycle time.

Mold temperature is a critical variable in compression molding. The higher it is, the faster the material cures. But if the material cures too fast, it will not have enough time to fill thin sections and far corners of the mold, and so a compromise must be reached. Material suppliers attempt to optimize the cure characteristics to provide minimum cycle times.

Although used occasionally in the laboratory, compression molding is impractical for thermoplastics, as the part has to be cooled before ejection from the mold. This requires alternate heating and cooling of a large mass of metal (the mold), which wastes much time and energy.

A variation of compression molding is known as *transfer molding.* Here, the material is melted in a separate transfer pot, from which it is squirted into the mold. This can give faster cycle times, and since no solid material is pushed around in the mold cavity itself, damage to delicate inserts (e.g., metal electrical contacts molded into a part) is minimized, mold wear is reduced, and greater ease in filling intricate molds and more uniform cures are obtained. Material cured in the transfer pot and sprue is wasted, however.

Reaction injection molding (RIM) (Fig. 19.3), sometimes called liquid injection molding (LIM), is a process developed fairly recently to mold large polyurethane (Example 4, Chapter 2, Q) articles directly from the starting chemicals with cycle times comparable to injection molding. Between shots, the highly reactive liquid polyol and isocyanate components are circulated from separate tanks through heat exchangers that regulate their temperatures. During the shot, the streams are pumped at about 2500 psi through nozzles in a mixing head, where they are impinged at high velocity and thoroughly mixed by the resulting turbulence. The mixed stream begins to react as it flows into the mold, where the polymerization reaction is completed. The mold is maintained at about 150°F, initially heating the reacting system but later removing the exothermic heat of polymerization as the reaction approaches completion. Since the unreacted stream is relatively low in viscosity (about like pancake syrup), unlike the molten polymer in an injection molding process, it flows easily to fill large molds with narrow clearances at relatively low pressures (100 psi, or so, in contrast to the thousands of pounds per square inch in injection molds). These low pressures

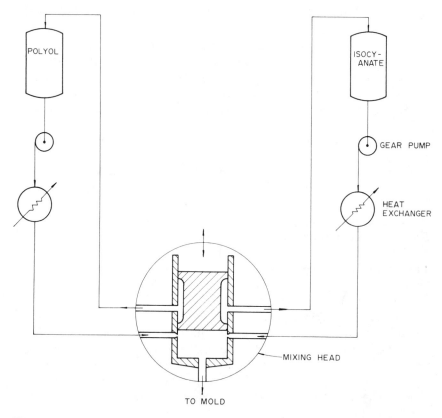

Figure 19.3   Reaction Injection Molding (RIM). The Kraus-Maffei mixing head is shown in the shot position. At the end of the shot, the plunger moves down, packing the mold, cleaning the mixing chamber, and recirculating the reactant streams back to their storage tanks.

allow the use of relatively inexpensive molds and low-force mold clamping systems (compared to injection molding).

The polyurethanes may be formulated to be flexible or rigid, solid or foamed. The major current application of RIM is to produce energy-absorbing front and rear panels for automobiles. Cycle times of 2 min or less are feasible for such large parts. The major current limitation on cycle time seems to result from the fact that polyurethanes are excellent adhesives. A mold-release agent has to be applied to the mold surfaces before each shot to keep the parts from sticking in the molds. Manufacturers are hard at work developing internal release agents to be added directly to one of the reactive components.

RIM has tremendous potential, particularly in the automotive industry, where the drive to cut weight is intense. It probably won't be long before RIM panels replace stamped steel in a variety of automative applications. Work is currently progressing in adapting the process to other reactive two-liquid-component systems, for example, epoxies (Chapter 2, Example 4, O), and to reinforced materials, in which short reinforcing fibers are incorporated in one or both of the reactants.

19.3   EXTRUSION

Thermoplastic items with a uniform cross-sectional area are formed by extrusion. This includes many familiar items such as pipe, hose and tubing, gaskets, wire and cable insulation, and sheeting. Molding powder is conveyed down an electrically or oil-heated barrel by a rotating screw. It melts as it proceeds down the barrel and is forced through a *die* that gives it its final shape (Fig. 19.4). Vented extruders incorporate a section in which a vacuum is applied to the melt to remove volatiles, such as traces of unreacted monomer, moisture, solvent from the polymerization process, or degradation products.

The design of extruders screws is an interesting and complex technical problem that has received considerable study.[10] Screws are optimized for the particular polymer being extruded. Basically, a screw consists of three sections: melting, compression, and metering. The function of the melting section is to convey the solid pellets forward from the hopper and convert them into molten polymer. Its analysis involves a combination of fluid and solid mechanics and heat transfer. The compression section, in which the depth of the screw flight decreases, is designed to compact and mix the molten polymer to provide a more or less homogeneous melt to the metering section, the function of which is to pump the molten polymer out through the die. This last section is well understood. Analysis of the metering section is an interesting application of the rheological principles discussed previously, as is much of die design. The determination of

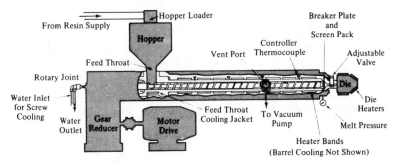

Figure 19.4   Vented Extruder (Modern Plastics Encyclopedia, McGraw-Hill, 1969–1970 ed.).

the die cross section needed to produce a desired product cross section (other than flat or circular) is still pretty much a trial-and-error process, however. Viscoelastic polymer melts swell upon emerging from the die (recovering stored elastic energy), and the degree of die swell cannot be reliably predicted.

In addition to die swell, as the extrusion rate is increased, the extrudate begins to exhibit roughness and then an irregular, severly distorted profile. This phenomenon is known as *melt fracture*. There is not yet general agreement as to its causes, but melt fracture can be minimized by increasing die length, smoothly tapering the entrance to the die, and raising the die temperature.

Extruders are normally specified by screw diameter and length-to-diameter ratio. Diameters range from 1 in. in laboratory or small production machines to a foot for machines used in the final pelletizing step of production operations. Typical L/D ratios seem to grow each year or so, with values now in the 20/1 to 36/1 range.

Single-screw extruders depend for their pumping ability on the drag flow of material between the rotating screw and stationary barrel. As a result, they are not positive-displacement pumps and tend to give a rather broad residence time distribution. Moreover, they are not particularly good mixing devices. Counter-rotating *twin-screw* extruders are true positive-displacement pumps, capable of generating the high pressures needed in certain profile extrusion applications. Corotating twin-screw extruders, though not positive displacement pumps, with propert screw design, can give excellent mixing and a narrow residence time distribution (subjecting all the material to essentially the same shear and temperature history). They are, therefore, used extensively in polymer compounding (mixing) operations, and to a certain extent as continuous polymerization reactors.

The heaters on extruders are needed mainly for startup. In steady-state operation, most or all of the heat for plasticizing the polymer is supplied by the drive motor through viscous energy dissipation. In fact, cooling through the barrel walls and/or screw center is sometimes necessary.

Most packaging film is produced by *blow extrusion* (Fig. 19.5). A thin-walled hollow cylinder is extruded vertically upward. Air is introduced to the interior of the cylinder, expanding it to a tube of film (less than 0.01-in. thickness). The tube is grasped between rolls at the top, preventing the escape of air and flattening it for subsequent slitting and windup on rolls. The expanded tube is rapidly chilled by a blast of air from a chill ring as it proceeds upward. In the case of polyethylene, this rapid chilling produces smaller crystallites and enhances film clarity.

19.4   BLOW MOLDING

A quick walk through a supermarket provides convincing proof of the economic

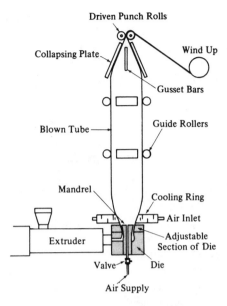

Figure 19.5   Blow Extrusion Film Line (Modern Plastics Encyclopedia, McGraw-Hill, 1969–1970 ed.).

importance of plastic bottles. They're nearly all made by blow molding. In one form of this process, *extrusion blow molding* (Fig. 19.6), a hollow cylindrical tube or *parison* is extruded downward. The parison is then clamped between halves of a water-cooled mold. The mold pinches off the bottom of the parison and forms the threads in the neck of the bottle. Compressed air expands the parison against the inner mold surfaces, and when the part has cooled sufficiently, the mold opens, the part is ejected, and the mold is returned to grab another parison.

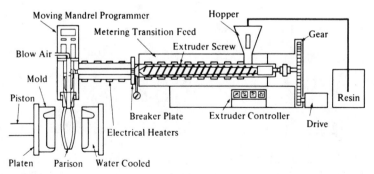

Figure 19.6   Extrusion Blow Molding (Modern Plastics Encyclopedia, McGraw-Hill, 1969–1970 ed.).

The rheological properties of the parison are important. If it sags too much before being grasped by the mold, the walls of the bottle will be too thin in places. Sag is minimized by using high-molecular-weight compounds with high viscosities. Many blow-molding extruders are equipped with *parison programmers*, which vary the orifice diameter as the parison is being extruded, providing any desired variation in wall thickness in the blown product.

Extrusion blow molding has passed well beyond the bottle-production stage. It is now being used to produce large items such as drums (to replace the familiar 55-gal steel drum) and truck, automobile, and recreational-vehicle gasoline tanks. In such metal replacement applications, blow-molded containers offer light weight and great design flexibility.

In *injection blow molding*, the parison is formed by injection molding, with the end closed and the threads molded on. A variation on this process known as *stretch* (or *orientation*) *blow molding* has recently begun to assume great commercial importance. In this process, parisons (sometimes called *preforms* in this case) are injection molded and allowed to cool to room temperature. They are reheated in a radiant-heat oven in which close control is excercised over the temperature profile of the parison. When introduced to the blow mold, the parison is normally a fraction of the length of the mold (Fig. 19.7), but before blowing, a rod rapidly stretches the parison to the full length of the mold, orienting the polymer molecules in the axial direction. This is followed by the blowing process, which imparts tangential (hoop) orientation. The resulting *biaxial* orientation improves the toughness, creep resistance, clarity, and barrier properties (resistance to permeation) of the bottle material. In this case, the frozen-in strains that are normally detrimental to thick injection-molded parts (Example 1) are deliberately introduced to thin-walled blown bottles with highly beneficial results. The biaxial orientation provided by stretch blow molding is absolutely essential to the success of the poly(ethylene terephthalate) soda-pop bottles that are rapidly penetrating the market.

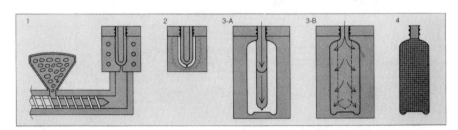

Figure 19.7   Stretch Blow Molding. (1) Parison injection molded. (2) Parison is reheated. (3-A) A rod stretches the parison, imparting axial orientation. (3-B) Air expands parison against mold walls, imparting tangential orientation. (4) Finished bottle ejected from mold. (*Modern Plastics*, *55*, 10, 22 (1978).)

19.5   ROTATIONAL, FLUIDIZED-BED, AND SLUSH MOLDING

Molding techniques have been developed to take advantage of the availability of finely powdered plastics, mainly polyethylene and nylons. In rotational molding, a charge of powder is introduced to a heated mold, which is then rotated about two mutually perpendicular axes. This distributes the powder over the inner mold surfaces, where it fuses. The mold is then cooled by compressed air or water sprays, opened, and the part is ejected. Rotational molding is capable of producing extremely irregular hollow objects. The molds are inexpensive, often simply sheet metal, because no elevated pressures are involved, and they can be heated in simple hot-air ovens. Thus capital outlay is relatively low. The process is in many ways competitive with extrusion blow molding for the production of large hollow items such as drums and gasolene tanks. Blow molding requires a much larger initial investment but is capable of higher production rates.

These two processes also provide a good example of how the processing operation, polymer, and finished part properties are often intimately connected. Blow molding can handle very high-molecular-weight linear polyethylenes, in fact often *requires* high-molecular-weight material to prevent excessive parison sag. Rotational molding, on the other hand, requires a low-molecular-weight resin because a low viscosity is needed to permit fusion of the powder under the influence of the low forces in a rotational mold. When subjected to stresses for long periods of time, particularly in the presence of certain liquids, linear polyethylene has a tendency to fail through stress cracking. It turns out that high-molecular-weight resins are much more resistant to stress cracking. Thus, although the parts might appear similar, those produced by extrusion blow molding will ordinarily have superior resistance to stress cracking. The difference can be narrowed or eliminated by using more material (thereby lowering stress for a given load) or by using a crosslinkable polyethylene in rotational molding. Either of these solutions increases material cost, however.

Polymer powders are also used in a process known as *fluidized-bed coating.* When a gas is passed up through a bed of particles, the bed expands and behaves much like a boiling liquid. When a heated object is dipped into a bed of fluidized particles, those that contact it fuse and coat its surface. Such *100%-solids* coating processes are increasing in importance because they eliminate the pollution often caused by solvent evaporation when ordinary paints are used.

Similar procedures have been in use for years with plastisols (Chapter 7). Liquid plastisol is poured into a heated female mold. The plastisol in contact with the mold surface solidifies, and the remainder is poured out for reuse. This *slush molding* process is used to produce objects such as doll's heads. Dipping a heated object into a liquid plastisol coats it with plasticized polymer. Vinyl-coated wire dishracks are familiar products of this process.

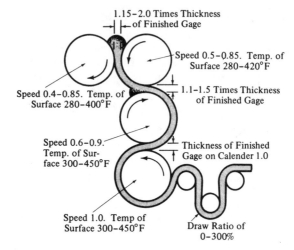

Figure 19.8   Inverted "L" Calender, Illustrating Process Variables for the Production of PVC Sheet (Modern Plastics Encyclopedia, McGraw-Hill, 1969–1970 ed.).

19.6   CALENDERING

Polymer sheet (greater than 0.01-in. thickness) may be produced either by extrusion through thin flat dies or by *calendering* (Fig. 19.8). Basically, a calender consists of a series of rotating heated rolls, between which the polymer compound (most often plasticized PVC) is squeezed into sheet form, the thickness of the sheet being determined by the clearance between the rolls. Commercial calenders may be very large (rolls up to 3 ft in diameter by 8 ft in length) and represent a big capital outlay, but they are capable of tremendous production rates (up to 100 yards/min). In addition to forming unsupported sheeting, the polymer may be laminated to a layer of fabric between two rolls to give a supported sheeting. The final rolls may also be embossed to impart a pattern to the sheet. Shower curtains, vinyl upholstery materials (Naugahyde), and vinyl floor tile are produced by calendering.

19.7   SHEET FORMING

Thermoplastic sheet is converted to a wide variety of finished articles by processes known generically as sheet forming. Although the details vary considerably, these processes all involve heating the sheet above its softening point and forcing it to conform to a mold (Fig. 19.9). In vacuum forming, for example,

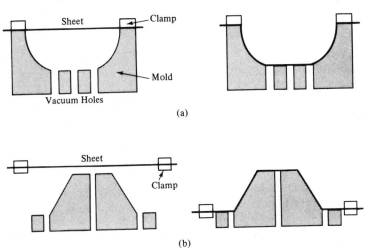

Figure 19.9   Sheet-forming Process: a. vacuum forming; b. drape forming.

the heat-plasticized sheet (either directly from an extruder or preheated in an oven) is drawn against the mold surface by the application of a vacuum from beneath the surface. Similarly, a positive pressure may be used to force the sheet against the mold surface, and where very deep draws are required, mechanical assists — plug forming — are used. In drape forming, the plasticized sheet is draped over a male mold, perhaps with a vacuum assist.

From a material standpoint, polymers used for sheet forming should have high "melt strengths," that is, high melt viscosities so that they do not draw down or thin out excessively or perhaps even tear in the forming operation. Thus high-molecular-weight resins are preferred. One of the great advantages of sheet forming is that relatively inexpensive molds are required since no high pressures are involved. Epoxy molds are often used because they can be easily cast to shape from a handmade pattern.

Among the many familiar items made by sheet forming are drinking cups, meat trays, cigarette packs, aircraft canopies, advertising displays, and lighting globes.

19.8   STAMPING

The plastics industry has long envied the ability to stamp sheet-metal parts with cycle times of a few seconds. Because of the high production rates possible with a stamping process (this is one important reason why most auto bodies are largely stamped steel), work is progressing with the aim of adapting such a process to forming thin-walled (0.1-in.) parts from glass-reinforced thermoplastic sheet (Azdel®). The sheet is heated in an infrared oven and then stamped in modified metal-forming equipment, with a dwell time of about 8 sec as the material cools in the mold. Because of the thin sections, cooling is rapid and cycle times as low as 15 to 20 sec can be achieved with automated feeding and part-removal systems.[13,14] So far, the process has been limited to rather simple shapes and shallow draws (crankcase oil pans, battery trays, bucket seat bottoms and backs, etc.), but because of its economic potential it should achieve more widespread use in the near future.

19.9   SOLUTION CASTING

Plastic sheet or film may be produced by dissolving the polymer in an appropriate solvent, spreading the viscous solution onto a polished surface, and evaporating the solvent. In the manufacture of photographic film base, the solution is spread with a "knife" onto a slowly rotating wheel about 2 ft wide and 20 ft in diameter. As the wheel revolves, heated air evaporates the solvent (which is recovered for reuse), and the dried film is stripped from the wheel before the casting point is again reached.

19.10   CASTING

The raw materials for many thermosetting polymers are available as low-molecular-weight liquids which are converted to solid or rubbery materials by the crosslinking reaction. These materials are easily cast to shape at atmospheric pressure in inexpensive molds. A good example of this process is the production of "carved wood" furniture panels from polyester resins. Only the original need actually be carved (which is the expensive step). A mold is made by pouring a room-temperature curing silicone rubber over the original. When cured, the rubber mold is stripped from the original and used to cast the polyester copies from a liquid unsaturated polyester (Chapter 2, Example 3). A free-radical initiator causes crosslinking to the final solid object, which reproduces faithfully the detail of the carved original at a fraction of the cost.

Such furniture components provide a good example of how the processing method of choice often depends on the production volume. They can also be injection molded from high-impact polystyrene, for example. Because the panels are usually large, they require a large and very expensive molding machine. Similarly, the mold, with its intricate carving and wood-grain detail in hardened steel, is very expensive to produce. Thus the capital outlay for injection-molded panels is huge, while the reverse is true for cast panels. Nevertheless, parts can be injection molded with cycle times under a minute and very little hand labor involved, and when the capital investment is amortized over a large number of panels, it becomes a small contribution to the total panel cost. On the other hand, the long cure times and hand labor required for the casting process make it prohibitively expensive for large production runs, but it is the preferred process for small numbers of parts.

Delicate electrical components are often encapsulated or "potted" by casting a thermosetting liquid resin (usually an epoxy) around them. The casting of acrylic sheet from monomer or syrup is described in Chapter 13.

Example 2. A polymer of divinyl benzene $H_2C=\overset{\overset{\displaystyle H}{|}}{C}-\langle\bigcirc\rangle-\overset{\overset{\displaystyle H}{|}}{C}=CH_2$ is to be used to make rotameter bobs (cylinders 1 in. in length by $\frac{1}{2}$ in. in diameter). What processing technique should be used?

Solution. The monomer is tetrafunctional (Chapter 2) and hence will be highly crosslinked in the polymerization reaction; therefore it cannot be softened by heat. The *only* suitable technique is casting directly from the monomer to final shape, or casting rods that can then be cut and machined to the final shape.

19.11   REINFORCED THERMOSET MOLDING

Many plastics, when used by themselves, do not possess enough mechanical strength for structural applications. When reinforced with high-modulus fibers, however, the composites have high strength-to-weight ratios and can be fabricated into a wide variety of complex shapes. Glass and other fibers are used extensively to reinforce thermosetting plastics, mostly polyesters and epoxies. FRPs (fiber-glass-reinforced plastics) are now used for just about all boats under 40 ft in length, truck cabs, low-production-volume automobile bodies, structural panels, aircraft components, and so on. Many such objects are fabricated by a process known as *hand layup*. The mold surface is often first sprayed with a pigmented but nonreinforced *gel coat* of the liquid resin to provide a smooth surface finish. The gel coat is followed up by successive layers of fiber glass, either in the form of woven cloth or random matting, impregnated with the liquid resin, which is then cured (crosslinked) to give the finished product. This process is tremendously versatile. The molds are relatively inexpensive because no pressure is required. Objects may be selectively reinforced by adding extra layers of material where desired. The major drawback of the process is the expense of the hand labor involved, which makes it uneconomical for large production volumes. For this reason it is giving way to a *sprayup* process. A special gun chops continuous fibers into approximately 1-in. lengths. The chopped fibers are combined with a stream of liquid resin and sprayed directly onto the mold surfaces. Although the random chopped fibers do not reinforce quite as well as woven cloth or random mat, the labor savings are substantial. Sprayed-on polyester–fiber glass backings are also applied to vacuum-formed polymethyl methacrylate sheets, thus combining a smooth, hard stain- and ultraviolet-resistant acrylic surface skin with the lightweight strength of a fiber-glass-reinforced plastic. Such composites are used for sinks, bathtubs, recreational vehicle bodies, etc.

In many objects, stresses are not uniformly distributed; so the reinforcing fibers may be arranged to support the stress most efficiently. In "fiber-glass" or "graphite" fishing rods, vaulting poles and golf club shafts, the fibers are arranged along the long axis to resist the bending stresses applied. The process of *filament winding* extends this principle to more complex structures. Continuous filaments of the reinforcing fiber are impregnated with liquid resin and then wound on a rotating *mandrel*. The winding pattern is designed to resist most efficiently the anticipated stress distribution. The range of the Polaris submarine ballistic missile was increased by several hundred miles by replacing the metallic rocket casing with a filament-wound reinforced plastic system. This technique is also used to produce tanks and pipe for the chemical process industries, and even gun barrels, by filament winding about a thin metal core that provides the necessary heat and abrasion resistance.

Pultrusion is used to produce continuous lengths of objects with a constant cross section, for example, structural beams. Continuous fibers (roving) and/or mat are impregnated by passing them through a tank of liquid resin and pulled slowly through a heated die of the desired cross section. The resin cures to a solid in the die.

19.12   FIBER SPINNING

The polymers used for synthetic fibers are similar, and in many cases identical, to those used as plastics, but for fibers, the processing operation must produce an essentially infinite length-to-diameter ratio. In all cases this is accomplished by forcing the plasticized polymer through a *spinnerette*, a plate in which a multiplicity of holes have been formed to produce the individual fibers, which are then twisted together to form a thread for subsequent weaving operations. The cross section of the spinnerette holes obviously has a lot to do with the fiber cross section, which in turn greatly influences the properties of the fiber. The three basic types of spinning operations differ mainly in the method of plasticizing and deplasticizing the polymer.

Melt spinning is basically an extrusion process. The polymer is plasticized by melting and pumped through the spinnerette. The fibers are usually solidified by a crosscurrent blast of air as they proceed to the drawing rolls. The drawing step stretches the fibres, orienting the molecules in the direction of stretch and inducing high degrees of crystallinity, a necessity for good fiber properties. Nylons are commonly melt spun.

In *dry spinning*, a solution of the polymer is forced through the spinnerette. As the fibers proceed downward to the drawing rolls, a countercurrent stream of warm air evaporates the solvent. In this process the cross section of the fiber is determined not only by the shape of the spinnerette holes, but also by the complex nature of the diffusion-controlled solvent evaporation process, because there is considerable shrinkage as the solvent evaporates. The acrylic fibers (Orlon, Creslan, Acrilan, etc.), mainly polyacrylonitrile, are produced by dry spinning.

Wet spinning is similar to dry spinning in that a polymer solution is forced through the spinnerette. Here, however, the solution strands pass directly into a liquid bath. The liquid might be a nonsolvent for the polymer, precipitating it from solution as the solvent diffuses outward into the nonsolvent. The bath might also contain a substance that precipitates a polymer fiber by chemically reacting with the dissolved material. Here again, the process as well as the spinnerette influences the fiber cross section. Rayon (regenerated cellulose) is a common example of a wet-spun fiber.

Example 3.   Suggest processing techniques for the manufacture of

a.   100,000 ft of plasticized PVC garden hose.

b.   50,000 polystyrene pocket combs.

c.   100,000 polyethylene detergent bottles.

d.   5000 phenolic (phenol-formaldehyde) TV knobs.

e.   Six souvenir paperweights of polymethyl methacrylate containing an old coin.

f.   A strip of chlorinated rubber (a linear, amorphous polymer) and white pigment (60:40), roughly 0.001 in. thick, 4 in. wide, running 20 miles down the center of a highway.

g.   1,000,000 polystyrene meat trays, 0.005 in. thick.

Solution

a.   Extrusion.

b.   Injection molding.

c.   Blow molding.

d.   Compression, transfer, or thermoset injection molding.

e.   Although these objects could be injection molded, the small number of articles required would make it uneconomic. They can easily be *cast* from the monomer.

f. Dissolve rubber in solvent, add pigment, and spray on highway. Evaporation of solvent leaves the desired strip.

g. Extrude sheet, and then vacuum-form trays.

19.13 COMPOUNDING

Polymers are almost always used in combination with other ingredients. These ingredients are discussed in subsequent chapters, but they must be combined with the polymer in a *compounding* operation.

Occasionally, if they don't interfere with the polymerization reaction, such ingredients may be incorporated at the monomer or low-molecular-weight polymer (e.g., unsaturated polyester, Chapter 2, Example 3) stage and carried through the polymerization and/or crosslinking reaction. In such cases viscosities are low enough to permit the use of standard mixing equipment. Similarly, powdered PVC and thermosets are compounded with other ingredients in the usual tumbling type of blending equipment.

Because of their extremely high melt viscosities, specialized equipment is usually needed to compound ingredients with high-molecular-weight thermoplastics, however. In general, high shear rates and large power inputs per unit

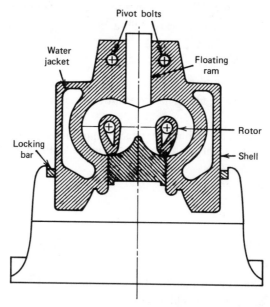

Figure 19.10   Banbury Mixer. (Arnold, L. K., *Introduction to Plastics*, The Iowa State University Press, Ames, Iowa (1968).)

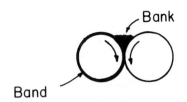

Figure 19.11   Two-roll Mill.

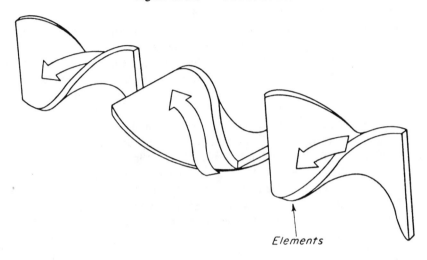

Elements

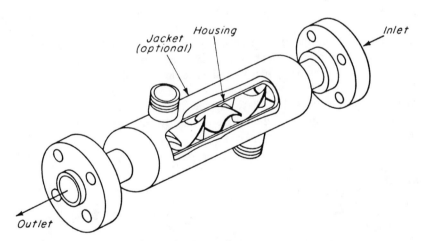

Figure 19.12   The Kenics *Static Mixer*®. (McCabe, W. L. and J..C. Smith, *Unit Operations of Chemical Engineering*, 3rd ed., McGraw-Hill, New York, 1976.)

volume of material are required to achieve a uniform and intimate dispersion of ingredients in the melt. Single- and twin-screw extruders (Section 19.3) are used extensively for continuous compounding. The latter, with screws often modified to incorporate special mixing sections, are better mixers and provide a narrower, more uniform distribution of residence times, while the former offer low cost and greater mechanical simplicity. *Intensive mixers*, such as the Banbury (Fig. 19.10), subject the material to high shear rates and large power inputs in a closed, heated chamber containing rotating, intermeshing blades. A *two-roll mill* (Fig. 19.11) generates high shear rates in a narrow nip between two heated rolls that counterrotate with slightly different velocities. In commercial mills, the rolls are about 1 ft in diameter by 3 ft in length. Once the polymer has banded on one of the rolls, ingredients are added to the bank between the rolls. The band is cut off the roll with a knife, rolled up, and fed back to the nip at right angles to its former direction. This is done several times to improve mixing along the length of the rolls. Despite extensive safety precautions, operators of two-roll mills often have n fingers ($n < 10$). To the author's knowledge, no systematic studies of the effect of the additional $(10 - n)$ ingredients on the properties of the compounded polymers have been reported.

Motionless mixers are a relatively recent development for continuous compounding.[15-17] One example of this type of device, the Kenics mixer (Fig. 19.12) consists of a series of alternating right- and left-hand helices that continuously divide the melt and cause it to rotate around its own hydraulic axis. Each element in a Kenics mixer breaks the stream into two parts. Therefore, an n-element mixer generates 2^n layers; so n does not have to be very large to achieve intimate mixing. The obvious advantage of this type of device is that it contains no moving parts (although the material must be pumped through it). It also requires relatively low power input per unit of material processed. Such mixers provide good radial mixing and relatively narrow residence-time distributions, despite the fact that flow is invariably laminar in the processing of polymer melts. They are also incorporated into extruders, between the screw and die, and into injection molding machines ahead of the nozzle to improve thermal homogeniety of the melt.

REFERENCES

1. N. M. Bikales, Ed., *Encyclopedia of Polymer Science and Technology*, Wiley, New York, 1971.

2. D. C. Miles and J. H. Briston, *Polymer Technology*, Chemical Publishing, New York, 1965, Part III.

3. *Modern Plastics Encyclopedia*, McGraw-Hill, New York (yearly).

4. F. Rodriguez, *Principles of Polymer Systems*, McGraw-Hill, New York, 1970, Chapter 12.

5. E. C. Bernhardt Ed., *Processing of Thermoplastic Materials*, Reinhold, New York, 1959.

6. C. D. Han, *Rheology in Polymer Processing*, Academic Press, New York, 1976.

7. J. M. McKelvey, *Polymer Processing*, Wiley, New York, 1962.

8. S. Middleman, *Fundamentals of Polymer Processing*, McGraw-Hill, New York, 1977.

9. J. R. A. Pearson, *Mechanical Principles of Polymer Melt Processing*, Pergamon Press, Oxford, 1961.

10. Z. Tadmor and I. Klein, *Engineering Principles of Plasticating Extrusion*, Van Nostrand Reinhold, New York, 1970.

11. Z. Tadmor and C. G. Gogos, *Principles of Polymer Processing*, Wiley-Interscience, New York, 1979.

12. I. I. Rubin, *Injection Molding. Theory and Practice*, Wiley-Interscience, New York, 1973.

13. S. Sikes, *Mod. Plast.*, **51**, 1, 70 (1974).

14. L. G. Ward, *Plast. Eng.*, **35**, 4, 47 (1979).

15. S. M. Skoblar, *Plast. Technol.*, Oct. 1974, p. 37.

16. N. R. Schott, B. Weinstein, and D. LaBombard, *Chem. Eng. Prog.*, **71**, 1, 54 (1975).

17. S. J. Chen, *Chem. Eng. Prog.*, **71**, 8, 80 (1975).

CHAPTER 20

Plastics

20.1   INTRODUCTION

As was mentioned in the introduction, there are five major applications for polymers: plastics, rubbers, synthetic fibers, surface finishes, and adhesives. Previous sections have dealt with the properties of the polymers themselves. While these properties are undoubtedly the most important in determining the ultimate application, polymers are rarely used in a chemically pure form; so in a discussion of the technology of polymers, it is necessary to mention, in addition to the properties of the polymers required for a given application, the nature and reasons for use of the many other materials often associated with the polymers. The following brief chapters will do this in turn for the five major applications.

Plastics are normally thought of as being polymer compounds possessing a degree of structural rigidity – in terms of the usual stress-strain test, a modulus on the order of 10^9 dynes/cm^2 or greater. The *molecular requirements* for a polymer to be used in a plastic compound are (a) if linear or branched, the polymer must be below its glass-transition temperature (if amorphous) and/or below its crystalline melting point (if crystallizable) at use temperature; otherwise (b) it must be crosslinked sufficiently to restrict molecular response essentially to straining of bond angles and lengths (e.g., ebonite or "hard" rubber).

20.2   MECHANICAL PROPERTIES OF PLASTICS

The engineering properties of commercial plastics vary considerably within the broad definition given above. Figure 20.1 sketches some representative stress—strain curves for three types of plastics. As discussed previously, for any given material, the quantitative nature of the curves depends markedly on the rate of strain and the temperature. In general, faster straining and lower temperatures lead to higher moduli (slopes) and smaller ultimate elongations.

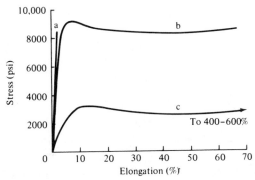

Figure 20.1   Typical stress-strain curves for plastics. a, rigid and brittle; b, rigid and tough; c, flexible and tough.

Plastics with stress–strain curves of the type a in Fig. 20.1 are *rigid* and *brittle*. The former term refers to the high initial modulus. The latter refers to the area under the stress–strain curve, which represents the energy per unit volume required to cause failure. These materials usually fail by catastrophic crack propagation at strains on the order of 2%. Since *hardness* correlates well with tensile modulus, it is another valuable property of this type of plastic. Examples of this class are polystyrene, polymethyl methacrylate, and most thermosets. Curve b represents *rigid* and *tough* materials, sometimes known as *engineering thermoplastics*. In addition to high modulus, tensile strength, and hardness, these materials undergo *ductile deformation* or drawing beyond the yield point, evidence of considerable molecular orientation before failure. This drawing confers toughness or *impact resistance*, that is, the ability to withstand shock loading without brittle failure. Examples of this class of materials are polycarbonates, cellulose esters, and nylons. Although the reasons for their toughness are not entirely clear, such plastics generally exhibit a secondary dynamic damping peak (Chapter 18) well below their use temperature. Curve c represents *flexible* and *tough* plastics, as typified by low- and medium-density polyethylenes. Here, the ductile deformation leading to very high ultimate elongations results from the conversion of low-crystallinity material to high-crystallinity material. The tensile samples "neck down," with the density, percent crystallinity, and modulus of the material in the neck being appreciably greater than those of the parent material. Despite their good toughness, these materials are limited in their structural applications by their low moduli and tensile strengths.

20.3   CONTENTS OF PLASTIC COMPOUNDS

In addition to the polymer itself, which is seldom used alone, plastics usually contain at least small amounts of one or more of the following additives.[1]

1. Reinforcing Agents.   The function of reinforcing agents is to enhance the structural properties of the compound, in particular, properties such as modulus (stiffness), strength, and the retention of these properties at higher temperatures. The use of long fibers to reinforce epoxy and polyester thermosets is mentioned in the previous chapter. While glass is by far the most common reinforcing fiber, largely due to its strength, ready availability, and relatively low cost, other, more exotic fibers are used where very high strength-to-weight ratios are required and cost is of secondary concern. Carbon or "graphite" fibers, made by the pyrolysis of pitch or polyacrylonitrile fibers, probably offer the best composite performance of all. While their high cost currently limits them to aerospace or small recreational applications (golf-club shafts, tennis-racket frames, etc.), they have been used successfully on an experimental basis to produce automotive springs and driveshafts, for example. As production increases, the cost will come down and applications will open up, particularly in the automotive industry, where the drive to save weight by replacing metal is intense. Aramid (*ar*omatic poly*amid*) fibers (Kevlar, Flexten) are also being used increasingly to reinforce plastics for premium-property applications.

Short ($\frac{1}{8}$- to $\frac{1}{2}$-in.) glass and other fibers are used extensively to reinforce thermoplastics. Addition of up to 40 wt% short glass fibers to polystyrene, nylon, polypropylene, and so on, provides a relatively inexpensive way of greatly improving the structural strength of the plastic and maintaining its strength to higher temperatures, as illustrated in Table 20.1. As might be expected, the addition of fibers makes processing somewhat more difficult and increases the wear on machine and mold, but the commercially available compounds are routinely injection molded.

Another example of reinforcement is the use of up to 20% dispersed rubber particles, on the order of 1 μm in diameter, to increase the *toughness* or *impact*

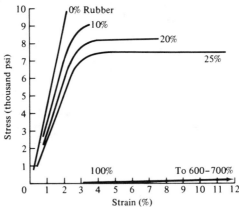

Figure 20.2   The influence of discrete, micron-sized rubber particles on the high-speed (133 in/in min) stress-strain curves of polystyrene.[2]

strength of a normally brittle plastic. Figure 20.2 illustrates the effect of adding increasing levels of a rubber to polystyrene in a high rate-of-strain (to simulate shock loading) tensile test. The addition of the rubber particles decreases the modulus (slope) and ultimate tensile strength, but introduces considerable ductile deformation (elongation beyond the "knee"), thereby increasing the area under the stress–strain curve. It also imports a secondary dynamic damping peak below use temperature. In effect, it converts a type a material (Fig. 20.1) to a type b material. For many applications, the improvement in impact strength considerably outweighs the slight decreases in modulus and tensile strength produced by the addition of rubber. *Impact modifiers* are rubbery additives that impart such toughness when blended with plastics.

Commercially important examples of these *rubber-toughened* plastics include the high-impact polystyrenes (HIPS), in which polystyrene is toughened with a polybutadiene rubber, and the ABS (acrylonitrile-butadiene-styrene) plastics, in which a polybutadiene or poly(butadiene-co-acrylonitrile) rubber toughens a poly(styrene (75%)-co-acrylonitrile (25%)) glassy phase. A more complete discussion of these two-phase polymer systems is available elsewhere.[2]

  2.   **Fillers.**   Fillers are low-cost particulate materials whose major function is to extend the polymer and thereby reduce the cost of the plastic compound. They have become increasingly important since the hydrocarbon crunch, which has roughly tripled the cost of most polymers.

One of the earliest examples of a filler is the wood flour (fine sawdust) long used in phenolics and other themosets. Calcium carbonate is used in a variety of plastics, and polypropylene is often filled with talc. Even water has been used to extend polyester casting compounds. Fillers may improve certain properties of the compound. They almost all reduce mold shrinkage and the thermal expansion coefficient. Mica and asbestos increase heat resistance. They also have some mechanical reinforcing effect and reduce warpage in molded parts. Hollow glass or phenolic microspheres reduce the density of the composite. Just about any filler whose hardness and modulus are greater than those of the polymer will increase the hardness and *initial* modulus of a mixture with that polymer, but, generally, elongation, ultimate strength, toughness, and processability suffer.

  3.   **Coupling Agents.**   In order that a reinforcing fiber such as glass be of maximum benefit (or a filler not be too detrimental to mechanical properties), stress must be efficiently transferred from the polymer to the reinforcing agent or filler. Unfortunately, most inorganics have hydrophilic surfaces, while the polymers are hydrophobic; so interfacial adhesion is often poor. This problem is exacerbated by the tendency of the surface of many inorganics (particularly glass) to adsorb water, which further degrades adhesion. Coupling agents are intended to help overcome these difficulties.

Currently, the most common coupling agents are silanes, with the general

Table 20.1. Some properties of short glass fiber reinforced thermoplastics (Fiberfil®)

	Nylon 66		Polystyrene		Polycarbonate	
	Unreinforced	Reinforced	Unreinforced	Reinforced	Unreinforced	Reinforced
Tensile strength, psi	11,800	20,000	8,500	14,000	9,000	20,000
Elongation, %	60	1.5	2.0	1.1	60–100	1.7
Tensile modulus, psi	400,000	1,000,000	400,000	1,100,000	320,000	1,300,000
Flexural strength, psi	11,500	28,000	11,000	17,500	12,000	26,000
Izod impact ($\frac{1}{4} \times \frac{1}{2}$-in. bar), ft-lb/in.	0.9	2.5	0.3	2.9	2.0	4.0
Heat-distortion point (264 psi), °F	150	498	190	220	280	300
Rockwell hardness	M79	M100	M70	M95	M70	M95

formula $YRSi(OR')_3$. The (OR') group reacts with the inorganic substrate, and the Y group reacts (or at least forms strong secondary bonds) with the polymer, thereby enhancing interfacial adhesion. The reaction sequence by which vinyl-triethoxysilane is believed to couple to a substrate containing hydroxyl functionality (e.g., glass) is shown below:

Step 1.   Hydrolysis (most likely with water adsorbed on the substrate):

$$H_2C=\overset{\overset{\displaystyle H}{|}}{C}-Si(OC_2H_5)_3 + 3H_2O \rightarrow H_2C=\overset{\overset{\displaystyle H}{|}}{C}-Si(OH)_3 + 3C_2H_5OH$$

Step 2.   Reaction with substrate:

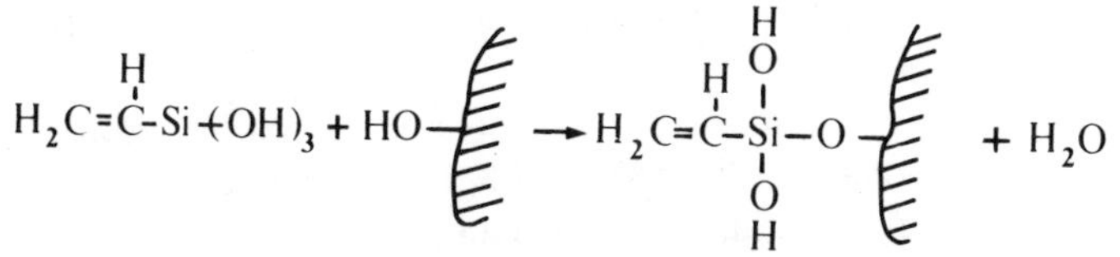

The "dangling double bonds" so introduced can then participate in the cure of an unsaturated polyester resin (Chapter 2, Example 3), covalently bonding the polymer to the surface.

A wide variety of silane coupling agents is available, with the functional groups optimized for use with various polymers and substrates. Titanate coupling agents perform in a similar fashion. Fillers and reinforcing fibers are supplied pretreated with coupling agent, or the agent may be added during compounding (in which case most of it presumably migrates to the interface).

In the case of fillers, coupling agents have an added benefit. By converting an otherwise hydrophilic surface to a hydrophobic surface, compatibility with the polymer is improved. This results in easier and more uniform dispersion of the filler in the polymer and, at a given filler loading, can reduce the viscosity substantially, thereby improving processability.

Graft and block copolymers often perform a coupling function in the case of the rubber-toughened plastics mentioned in the discussion of reinforcing agents. In the manufacture of high-impact polystyrene, for example, the polybutadiene is dissolved in the polymerizing styrene monomer. As a result, some of the styrene is grafted to the rubber. The grafted polystyrene chains are physically compatible with the surrounding homopolystyrene and are chemically bound to the rubber, thereby enhancing adhesion between the dispersed polybutadiene particles and the polystyrene matrix. The graft copolymer is thought to be essential for effective impact enhancement in this particular system, because simple mechanical blends of polystyrene and polybutadiene of similar composition do not show much impact enhancement. However, minor amounts of styrene–butadiene *block* copolymers *will* improve the compatibility and therefore the impact strength of such blends, much as a soap emulsifies oil in water.

4. Stabilizers. Most polymers are susceptible to one or more forms of degradation, usually as a result of environmental exposure to oxygen or ultraviolet radiation, or to high temperatures during processing operations. Stabilizers inhibit degradation of the polymer. In the case of polyvinyl chloride, for example, the major product of thermal degradation during processing is HCl, which catalyzes further degradation. Compounds that react with the HCl to form stable products, such as metal oxides, are used as stabilizers. Oxidative degradation of polymers is thought to take place by a free-radical mechanism involving crosslinking and/or chain scission initiated by free radicals from peroxides formed in the initial oxidation step. Similarly, these reactions can be initiated by free radicals produced by ultraviolet radiation. Stabilizers against such reactions are generally quinone-type organics which are effective free-radical scavengers.

5. Pigments. Plastics are often colored by the addition of pigments — finely divided solids. If the polymer is itself transparent, a pigment imparts opacity. Titanium dioxide is a common pigment where a brilliant, opaque white is desired. Sometimes pigments perform other functions. For example, calcium carbonate acts as both a filler and a pigment in many plastics, and carbon black is often a stabilizer as well as a pigment. It prevents degradation by absorbing and preventing the penetration of ultraviolet light beyond the surface of the article.

6. Dyes. Dyes are colored organic chemicals that dissolve in the polymer to produce a transparent compound (assuming the polymer is transparent to begin with). Some thermoplastics, though transparent, develop a slight yellow tinge from minor degradation during processing which causes selective absorption of light toward the blue end of the spectrum. The yellow can largely be "canceled out" by the addition of a blue dye which reduces the transmission of yellowish wavelengths. This technique unavoidably results in a slight lowering of the total light transmission.

7. Plasticizers. The addition of a relatively low-molecular-weight organic plasticizer to a normally glassy polymer will progressively reduce its modulus by bringing the compound's glass-transition temperature down closer to use temperature. The level at which the modulus is reduced to the point where the compound is considered an elastomer rather than a plastic is somewhat arbitrary, but plastics sometimes contain small amounts of a plasticizer to reduce brittleness.

8. Lubricants. *External* lubricants are low-molecular-weight organics that are relatively insoluble in the polymer and "plate out" or migrate to the surface of the compound and form a slippery coating during processing operations. They produce smoother extrudates and molded articles, and minimize sticking in the mold by acting as mold release agents. Where subsequent painting or printing of the surface is required, they can reduce adhesion, and so must be used with care. Stearic acid and its metal salts are common external lubricants.

Internal lubricants are soluble in the polymer and ease processing by lowering the compound's viscosity.

9. Processing Aids. Various compounds (often a second polymer) are used to modify the rheological properties of a polymer during processing to provide higher outputs, better surface finish, and easier handling in general. Low-molecular-weight polyethylene is added to PVC to improve extrusion behaviour.

10. Curing Agents. These are chemicals whose function is to produce a crosslinked, thermosetting plastic from an initially linear or branched polymer. A vinyl monomer such as styrene, a free-radical initiator, and sometimes a *promoter* (e.g., cobalt naphthenate — to speed up the reaction) are dissolved in a low-molecular-weight unsaturated polyester resin and crosslink it by an ordinary addition mechanism involving the double bonds in the polyester (Chapter 2, Example 3).

The polymer for a *two-stage* thermosetting compound is deliberately produced with a stoichiometric shortage of one of the reagents to give a highly branched, but not yet crosslinked structure, which is later cured in the mold with a curing agent. An "A-stage" phenolic polymer is produced by reacting phenol and formaldehyde in about a 1.25/1 mole ratio (a molal excess of formaldehyde is required for crosslinking). This still-thermoplastic product (a *novolac*) is compounded with fillers, pigments, reinforcing agents, and so on, and a curing agent, hexamethylenetetramine, to give a "B-stage" resin. The curing agent decomposes in the presence of moisture (a product of the condensation reaction) upon heating in the mold, giving the additional formaldehyde required for crosslinking and ammonia, which acts as a basic catalyst for the reaction.

| Hexamethylenetetramine | Water | Formaldehyde | Ammonia |

$$+ \; 6H_2O \longrightarrow 6 \; C{=}O \; + \; 4 NH_3$$

Single-stage resins (resoles) are made using the final desired reactant ratio, about 1/1.5 phenol/formaldehyde, but stopping the reaction short of crosslinking. The reaction is completed in the mold without the necessity of a curing agent.

Free-radical initiators (usually organic peroxides) are used as curing agents for saturated thermoplastic polymers. The curing agent must be compounded with the polymer at temperatures low enough to prevent appreciable decomposition to free radicals. The compound is then heated in the mold, producing

$$R \cdot R \longrightarrow 2R \cdot$$

Initiator　　　　　　Heat　　Free radical
(curing agent)

$$R \cdot \; + \; \sim\!\!\!\sim \underset{\underset{H}{\overset{\displaystyle H}{|}}}{\overset{\displaystyle H}{\underset{|}{C}}}-\underset{\overset{\displaystyle \cdot\cdot}{}}{C}-\underset{\overset{\displaystyle H}{|}}{\overset{\displaystyle H}{\underset{|}{C}}}\!\!\!\sim\!\!\!\sim \longrightarrow \sim\!\!\!\sim C\text{-}C\text{-}C \sim\!\!\!\sim \; + \; R:H$$

Polymer chain

Crosslinked chains

free radicals that abstract protons from the polymer, leaving unshared electrons on the chains which combine to form crosslinks.

Sometimes sulfur or multifunctional vinyl monomers are added to improve cross-linking efficiency by helping to "bridge the gap" (i.e., produce longer crosslinks) between the chains. In this manner, polyethylene is crosslinked with dicumyl peroxide, converting it from a thermoplastic to a thermoset with much greater heat resistance and resistance to stress cracking and abrasion. It can also tolerate much higher levels of carbon black filler without becoming excessively brittle.

　　11. Blowing Agents. Foamed plastics must contain a material that generates a gas to produce foaming. There are *chemical blowing agents* (CBAs) that generate gas through a chemical reaction and physical blowing agents, inert but volatile chemicals that are dissolved in the polymer or its precursors and simply vaporize upon heating or a reduction in pressure. As an example, polyurethanes (Chapter 2, Example 4, Q) may be "water blown" by the reaction of excess diisocyanate with water, generating CO_2,

$$O{=}C{=}N\text{-}R\text{-}N{=}C{=}O + 2H_2O \rightarrow R(NH_2)_2 + 2CO_2$$

Diisocyanate　　　　　　　　　Diamine

or, alternately, foamed with a chlorofluorocarbon physical blowing agent, usually CCl_3F. The latter, because of its low thermal conductivity, makes good insulating foams but is in imminent danger of being banned because of the possible threat it poses to atmospheric ozone.

CBAs must be compounded with a polymer below their decomposition temperatures and then decompose to generate the gas at temperatures low enough to prevent degradation of the polymer. A variety of CBAs is available, covering a range of decomposition temperatures, for use with most common plastics.[3,4]

In the manufacture of polystyrene foam molding beads, pentane is added to the monomer in a suspension polymerization or absorbed by the beads afterward. When the beads are placed in a mold and heated (usually with steam), the pentane volatilizes, expanding the beads against each other and the mold walls. These beads are used in many familiar applications, such as drinking cups, picnic coolers, and packaging supports.

12. Flame Retardants.[5-7]   Most synthetic polymers, being composed largely of carbon and hydrogen, are flammable (although no more so than the natural materials, e.g., wood and cotton, which they often replace – a fact sometimes overlooked). Plastics are increasingly being compounded with flame retardants to reduce their flammability. The most common flame-retarding additives for plastics contain large proportions of chlorine or bromine. These elements are believed to quench the free-radical flame propagation reactions. They may be compounds that are simply mixed with the plastic, for example, decabromodiphenyl ether, or they may be reactive monomers that become part of the polymer. Examples of the latter include tetrabromobisphenol-A substituted for some of the normal bisphenol-A in epoxies (Chapter 2, Example 4, O) or polycarbonates (Chapter 2, Example 4, N) and tetrabromophthalic anhydride in polyesters (Chapter 2, Example 3). Antimony trioxide is often used in synergistic combination with halogenated compounds. Organic phosphates are thought to function as flame retardants by forming a char that acts as a barrier to the flame. A compound such as tris(2,3-dibromopropyl) phosphate combines halogen and phosphate. Polyvinyl chloride, because of its chlorine content, is inherently flame resistant, but if it is to maintain this valuable property in plasticized form, it must be compounded with halogenated or phosphate plasticizers. Hydrated alumina is a particulate filler that contains 35% water of hydration, the evaporation of which absorbs energy and inhibits flame spread.

13. Miscellaneous.   Various other materials are added to plastics to provide certain end-use properties. For example, where a polymer will be subjected to a warm, humid environment (vinyl shower curtains, silicone-rubber bathtub caulks, etc.), it often contains a *biocide* to inhibit the growth of mold, mildew, and fungus. These biocides are generally organic copper, mercury, or tin compounds. Fatty acid amines are added to compounds for use in bottles and phonograph records as antistatic agents. These chemicals, because of their

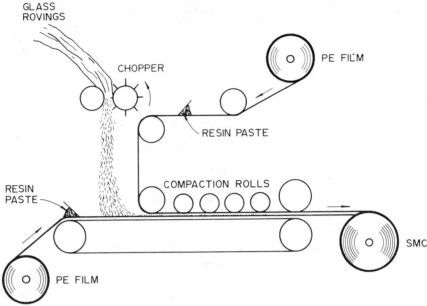

Figure 20.3   Sheet molding compound line.

limited compatibility with the polymer, migrate to the surface of the article and, because of their polarity, attract moisture from the atmosphere. The moisture bleeds off static electricity charges, which otherwise would attract considerable dust over a period of time. These same fatty acid amines also function as *slip* or *antiblock* agents, which prevent layers of plastic film and sheet from sticking to each other.

20.4   SHEET MOLDING COMPOUND

The subject of *sheet molding compound* (*SMC*) is introduced here because (a) they are commercially important, (b) they make use of many of the additives discussed in the previous section, and (c) they don't seem to fit logically anywhere else. While the details vary considerably, SMCs consist roughly of a third each of unsaturated polyester resin (Chapter 2, Example 3), filler (usually $CaCO_3$), and chopped-glass reinforcing fiber. The liquid polyester and filler are compounded in a high-shear mixer along with minor amounts of a curing agent (usually an organic peroxide), a thickener (MgO or CaO), an external lubricant as a mold-release agent (zinc stearate), and sometimes a low-shrink additive. The material leaves the mixer looking (but not smelling) for all the world like pancake batter and is spread continuously onto layers of polyethylene film (Fig. 20.3). Chopped glass fibers (about 1 in. long) are sprinkled on, the layers are combined, and the sheet is kneaded by rollers to wet out the glass. The sheet is then rolled up and allowed to age, during which time it thickens to the consistency of cardboard. This thickening is believed to involve interaction of the MgO or CaO with the carboxylic acid groups on the polyester, with water playing a role also, but *it is not* a true polymerization. Before use, the sheets are cut to size and the polyethylene stripped off. They are then placed in a heated compression-type mold (several layers may be stacked for greater thicknesses) where true cure (crosslinking) of the polyester takes place.

The choice of a curing agent for an SMC involves a compromise between an active initiator that gives quick cures but, conversely, imparts a short shelf life to the stored SMC and a less active initiator with a longer shelf life but slower cure. A drawback of early SMCs was the wavy surface and sometimes visible glass-fiber texture produced by shrinkage of the polyester during cure. This necessitated hand finishing operations, which are not at all popular in the automotive industry. The low-shrink additive that counteracts this is usually a second linear polymer dissolved in the polyester, which precipitates out as a micron-sized dispersed phase as the polyester cures. Why this should reduce shrinkage is not at all clear, but if the low-shrink additive is a rubber, it often acts as a toughening agent as well (see Reinforcing Agents).

Sheet molding compounds provide a high strength-to-weight ratio, with low cost and easy, rapid processing. They are used for automotive body parts, such as front ends and fender extensions, and for other structural parts, such as business machine and air-conditioner housings. They (along with RIM polyurethanes, injection-molded thermoplastic foams, and stamped sheet) are in the thick of the competition for increased use in auto body panels.

Bulk molding compounds (*BMCs*) are similar in composition to SMCs, with somewhat shorter fibers ($\frac{1}{4}-\frac{1}{2}$ in.), and perhaps less glass (15%) and more filler. They are produced in chunk form and compression molded like other thermosetting compounds.

REFERENCES

1. R. B. Seymour, Ed., *State of the Art*, Vol. 1: *New Developments*, Vol. 2: Additives for Plastics, Academic, New York, 1978.
2. S. L. Rosen, *Trans. N. Y. Acad. Sci.*, 35, 6, 480 (1973).
3. R. L. Heck, III, *Plast. Compd.*, Nov./Dec. 1978, p. 52.
4. R. S. Hallas, *Plast. Eng.*, Dec. 1977, p. 17.
5. L. H. Schongar, *Plast. Compd.*, May/June 1978, p. 44.
6. W. P. Blake, *Plast. Compd.*, July/August 1978, p. 26.
7. R. C. Nametz, *Plast. Compd.*, Jan./Feb. 1979, p. 31.

CHAPTER 21

Rubbers

21.1 INTRODUCTION

A rubber is generally defined as a material that can be stretched to at least twice its original length and that will retract rapidly and forcibly to substantially its original dimensions on release of the force. An elastomer is a rubberlike material from the standpoint of modulus but one that has limited extensibility and incomplete retraction. The most common example is highly plasticized polyvinyl chloride.

From a molecular standpoint, a rubber must be (a) a high polymer, since rubber elasticity is due mainly to the uncoiling and coiling of long chains (Chapter 14). In order for the molecules to be able to coil and uncoil freely, (b) the polymer must be above its glass transition temperature at use temperature. Furthermore, (c) the polymer must be amorphous in its unstretched state, as crystallinity would hinder the molecular motion necessary for rubber elasticity. Until fairly recently, an additional requirement was (d) that the polymer be crosslinked. If this were not so, the chains would slip past one another under stress (viscous flow) and recovery would be incomplete.

21.2 THERMOPLASTIC ELASTOMERS

The introduction of so-called thermoplastic elastomers (actually rubbers) has got around the requirement for crosslinking in the strictest sense of the term — covalent bonding between chains. The most common thermoplastic elastomers are styrene–butadiene–styrene (SBS) block copolymers, produced in solution by anionic polymerization (Chapter 11, Example 1). Since most polymer pairs are mutually incompatible (insoluble), largely because of the very low entropies of solution (Chapter 7), the polystyrene chain ends aggregate together in microscopic domains. These polystyrene domains, being below their glass-transition temperature $(100°C)$ at normal use temperatures, are rigid and act to tie together the long, flexible polybutadiene segments (above their T_g) as do ordinary crosslinks. Unlike the usual covalent crosslinks, however, the polystyrene domains melt above the T_g of polystyrene, and the polymer behaves as a true thermoplastic. Thus thermoplastic elastomers do not have to be vulcanized and can be processed by the usual economical thermoplastic techniques.

21.3 CONTENTS OF RUBBER COMPOUNDS

Natural rubber and, with a few exceptions, the many synthetic rubber polymers commercially available are unsaturated; that is, they contain double bonds that provide sites for the vulcanization (crosslinking) reaction. The polymers are mostly linear or branched, but many contain also (either intentionally or unintentionally) quantities of gel (crosslinked particles) before vulcanization. This can have a profound effect on processing properties. Also included in rubber compounds are

1. Reinforcing Agents.   Most rubber compounds that are intended to develop a reasonable tensile strength and abrasion and tear resistance will contain up to 50 phr (parts per hundred parts rubber by weight) or more of a reinforcing agent, nearly always a carbon black. The use of carbon black in rubbers is quite different than in plastics, where it is strictly a pigment and limited to much lower loadings. Carbon blacks are not simply carbon. Basic blacks have hydroxyl groups at the particle surfaces, and acid blacks have carboxylic acid functionality. It has been shown that the rubber polymer forms strong secondary and primary covalent bonds with the carbon black, which accounts for its reinforcing ability. There is a wide variety of carbon blacks available. In addition to chemical functionality, they differ in such factors as particle size, degree of aggregation, and surface area, and different types of rubber polymers require particular kinds of black for optimum reinforcement. Silicone rubbers are sometimes reinforced with finely divided silica (SiO_2).

Natural rubber (*cis*-1,4 polyisoprene) and its synthetic counterpart and butyl rubber are among the few rubber polymers that can develop reasonable mechanical strength without reinforcement. This is because they crystallize with molecular orientation at high elongations, and the crystallites function as a reinforcing agent, causing the sharp upturn in the stress–strain curves at high elongations shown in Fig. 21.2.[1]

2. Fillers.   As with plastics, the function of fillers in rubber is mainly to reduce the cost of the compound. The most common rubber fillers are finely divided inorganics such as $CaCO_3$. The addition of such high-modulus fillers

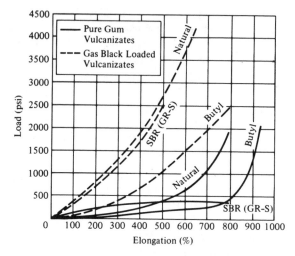

Figure 21.1   Stress–strain curves for filled and unfilled rubber vulcanizates (1). From Principles of High-Polymer Theory and Practice by A. X. Schmidt and C. A. Marlies. Copyright 1948 by the McGraw-Hill Book Company, Inc. Used with permission of McGraw-Hill Book Company.

raises the modulus of (stiffens) the compound, of course, and too much will cause a loss of rubbery properties, all other things being equal. Carbon blacks sometimes perform as fillers, as well as reinforcing agents, in rubber compounds.

3. Extending Oils.   Hydrocarbon oils are often used in rubber compounds. Their function is twofold. First, they plasticize the polymer, making it softer and easier to process. This is particularly important with very high-molecular-weight polymers. Second, since they are usually cheaper than the rubber polymer, they act like fillers in reducing the cost of the compound. Extending oils are available in various degrees of aromaticity, and the properties of the compound will depend on the type of oil in relation to the polymer, as well as on oil level.

Carbon black and extending oils have opposite effects on the modulus of a rubber compound. One trick for producing low-cost compounds from synthetic polymers is to polymerize to higher-than-normal molecular weight. The modulus is cut down by the addition of an extending oil and then brought back up to the desired level by the addition of large amounts of black. Projecting figures showing the increasing levels of black and oil in rubber compounds, it appears that there soon won't be any polymer left. Seriously, there is a limit to this technique, as other properties soon are degraded excessively.

4. Vulcanizing or Curing Systems.   The sole function of vulcanizing or curing systems is to crosslink the polymer. The most common curing systems are based on sulfur. While sulfur alone will cure unsaturated rubbers with heating, the process is slow and inefficient in its use of the sulfur. The mechanisms of sulfur curing are not well understood but are thought to include (among other things) the formation of sulfide or disulfide links between chains and the abstraction of protons from adjacent chains to form H_2S, with the chains cross-linking at the remaining unshared electrons.

To speed up the vulcanization process, *accelerators* are generally used. These are usually complex sulfur-containing organic compounds, often of proprietary composition. *Promoters* or *activators* are used to improve the cure still further. Zinc oxide is a common example, particularly in conjunction with stearic acid.

Nonsulfur cures are used occasionally. Free-radical initiators will provide crosslinks, as discussed in the previous chapter, and zinc oxide will crosslink polymers containing chlorine.

5. Antioxidants or Stabilizers.   Antioxidants or stabilizers are particularly important with natural rubber and the many synthetic rubbers that contain a major proportion of butadiene or isoprene. These polymers are highly unsaturated, and the double bonds are extremely susceptible to attack by oxygen and ozone, resulting in embrittlement, cracking, and general degradation. Since the degradation reactions take place by free-radical mechanisms, antioxidants are compounds that scavenge free radicals.

6. Pigments.   Where great mechanical strength is not required, and thus carbon black is not used, rubbers can be colored with pigments, as can plastics.

A typical tire-tread formulation is shown in Table 21.1. Notice that the final compound is less than 60% polymer.

Table 21.1[2]   Tire-tread formulation

Ingredient	Parts by weight
GR-S 1000 (75/25 butadiene/styrene emulsion copolymer)	100
HAF black	50
Zinc oxide ⎫ (promoters)	5
Stearic acid ⎭	3
Sulfur	2
Santocure (accelerator)	0.75
Circosol 2XH (extending oil)	10

21.4   RUBBER COMPOUNDING

The ingredients above must be *compounded* with the rubber polymer to produce the final rubber compound for molding, extrusion, and so on. This is usually done in either *two-roll* or *Banbury* mills (Section 19.13). Both devices are driven by relatively large electric motors and put a lot of energy per unit time and volume into the polymer. This energy input serves two purposes. First, it breaks down the polymer and reduces its "nerve" to the point where it can be easily compounded and processed. Nerve is a term used to describe the difficult-to-handle highly elastic response caused by too high a molecular weight. This mastication step mechanically degrades the polymer, lowering its molecular weight and therefore increasing viscous response. Premastication is particularly important with natural rubber, since no control can be exercised over its molecular weight in the polymerization step. Second, the high energy input breaks up and disperses the compounding ingredients evenly throughout the polymer.

REFERENCES

1.   A. X. Schmidt and C. A. Marlies, *Principles of High Polymer Theory and Practice*, McGraw Hill, New York, p. 537.
2.   G. S. Garvin, *Polymer Processes*, C. E. Schildknecht, Ed., Interscience, New York,

CHAPTER 22

Synthetic Fibers

22.1 INTRODUCTION

While many of the polymers used for synthetic fibers are identical to those in plastics, the two industries grew up separately, with completely different terminologies, testing procedures, and so on. Many of the requirements for fabrics are stated in nonquantitative terms such as "hand" and "drape" which are difficult to relate to normal physical property measurements, but which can be critical from the standpoint of consumer acceptance and, therefore, commercial success of a fiber.

A fiber is often defined as an object with a length-to-diameter ratio of at least 100. Synthetic fibers are spun in the form of continuous *filaments* but may be chopped up to much shorter *staple*, which is then twisted into thread before weaving. Natural fibers, with the exception of silk, are initially in staple form. The thickness of a fiber is most commonly expressed in terms of *denier*, which is the weight in grams of a 9000-m length of the fiber. Stresses and tensile strengths are reported in terms of *tenacity*, with units of grams per denier.

22.2 FIBER PROCESSING

Synthetic fibers are pretty much unoriented as they emerge from the spinning operation. To develop the tensile strengths and moduli necessary for textile fibers, they must be drawn (stretched) to orient the molecules along the fiber axis and develop high degrees of crystallinity. All successful fiber-forming polymers are crystallizable, and so from a molecular standpoint, the polymer must have polar groups between which strong hydrogen bonding holds the chains in a crystal lattice (e.g., polyacrylonitrile, nylons) or be sufficiently regular to pack closely in a lattice held together by dispersion forces (e.g., isotactic polypropylene).

319

22.3   DYEING

The dyeing of fibers is a complex art in itself. A successful dye must either form strong secondary bonds to polar groups on the polymer, or react to form covalent bonds with functional groups on the polymer. Furthermore, since the fibers are dyed after spinning, the dye must penetrate the fiber, diffusing into it from the dye bath. The size of the dye molecule is such that it cannot penetrate crystalline areas of the polymer; so it is mainly the amorphous regions that are dyed. This often conflicts with the requirement of high crystallinity. The chains of poly-acrylonitrile, for example, while possessing the necessary polar sites for dye attachment in abundance, are so strongly bound to each other that it is difficult for the dye to penetrate. For this reason, acrylic fibers usually contain minor amounts of plasticizing comonomers to enhance dye penetration. Nonpolar, nonreactive fibers such as polypropylene, on the other hand, have no sites to which the dye can bond even if it could penetrate. This was long a problem with polypropylene fibers and was overcome by incorporating a finely divided solid pigment in the polymer before melt spinning. Copolymers of propylene and a monomer with dye-accepting sites are now available.

22.4   EFFECTS OF HEAT AND MOISTURE

The polarity of the polymer also directly influences its degree of water absorption. Other things being equal, the more polar the polymer, the higher its equilibrium moisture content under any given conditions of humidity. As with dyes, however, moisture content is reduced by strong interchain bonding. The moisture content exerts a strong influence on the feel and comfort of fibers. Hydrophobic fibers tend to have a "clammy" feel in clothing and can build up static electricity charges. Perhaps the most important effect of moisture on polar polymers is as a plasticizer. Since fiber-forming polymers are linear, heat is also a plasticizer. This explains why suits wrinkle on hot, humid days, and why the wrinkles can be removed by steam pressing. The increasingly popular "wash-and-wear" and "permanent-press" fabrics are produced by operations that crosslink the fibers by reacting with functional groups on the chains, such as the hydroxyls on cellulose. The more hydrophobic polymers are inherently more wrinkle resistant because they are not plasticized by water. Wash-and-wear shirts, therefore, usually are made of blends of polyethylene terephthalate and cotton, about 65%/35%.

CHAPTER 23

Surface Finishes

23.1 INTRODUCTION

Nearly all surface finishes and coatings, with the exception of ceramic types for high-temperature applications, are based on a polymer film of some sort. They account for the use of a lot of polymer, but determining just how much and which polymers is not easy, because most formulations are proprietary, and production figures do not always separate polymer and nonpolymer components. The next section describes the five traditional types of surface finishes and the role played by polymers in each. Subsequent sections discuss more recent developments in the field.

23.2 TRADITIONAL TYPES OF SURFACE FINISHES

1. Lacquers. A lacquer consists of a polymer solution to which a pigment has been added. The film is formed simply by evaporation of the solvent, leaving the pigment trapped in the polymer film. Since no chemical change occurs in the polymer, it retains its original solubility characteristics. Hence the major drawback to lacquers is their poor resistance to organic solvents. Much of the technology of lacquers involves the development of polymers that form tougher, more adhesive, and more stable films and the choice of solvent systems that provide the optimum viscosity for application. The volatility of the solvent system is important, too. If it is too high, it will evaporate before the film has had a chance to "level," leaving brush marks, or an "orange-peel" or rough surface when sprayed. If too low, the coating will "sag" excessively after application.

A wide variety of polymers is used for lacquers. Newer systems favor acrylic polymers for their superior chemical stability. Acrylic lacquers may also incorporate a curing agent, for example, hexamethoxymelamine, which crosslinks the film in a subsequent baking operation. Crosslinking toughens the film and overcomes the major drawback of lacquers by making the film insoluble.

The term "spirit varnish" is an old and imprecise name for a lacquer in which the solvent is alcohol, for example, shellac. Since the polymers are soluble in a highly polar solvent, spirit varnishes have rather poor water resistance.

2.   **Oil Paints.**   Oil paints are popular and widely used finishes that consist of a suspension of pigment in a *drying oil*, an ester of glycerine with an unsaturated fatty acid such as linseed oil. The film is formed by a reaction involving atmospheric oxygen that polymerizes and crosslinks the drying oil through its double bonds. Sometimes an inert solvent (mineral spirits, turpentine) is added to control viscosity, and catalysts, for example, cobalt naphthenate, are used to promote the crosslinking reaction. Oil paints, once cured, are no longer soluble, although they may be softened considerably by appropriate solvents (as used in paint removers).

3.   **Oil Varnish or Varnish.**   Varnishes are coatings that consist of a polymer, either natural or synthetic, dissolved in a drying oil, with perhaps an inert solvent to control viscosity and a catalyst to promote the crosslinking reaction with oxygen. When cured, they produce a clear, tough, solvent-resistant film.

4.   **Enamel.**   An enamel is a pigmented oil varnish. It is much like an oil paint, except that the added polymer, replacing some of the drying oil, provides a tougher, glossier film.

5.   **Latex Paints.**   Latex paints are versatile finishes that are well on their way to replacing oil paints and enamels for home use because of their quick drying, low odor, and water cleanup properties. Basically, they are polymer latexes, produced by emulsion polymerization (Chapter 13) to which pigments and rheological-control agents have been added.

The film is formed by coalescence of the polymer particles upon evaporation of the water. The polymer itself is not water soluble. Although they are sometimes termed "water-soluble" paints, this is a serious misnomer, as is well known to anyone who has tried to clean a brush after the water has evaporated. As long as the individual latex particles have not coalesced, they are water-*dispersable.*

In order for the particles to coalesce to form a film when the water evaporates, the polymer must be deformable under the action of surface-tension forces. Thus, polymers for latex paints must be near or above their glass-transition temperatures at use temperature. This formerly resulted in a rather soft paint film, which inherently lacked "scrubbability." Reasonable resistance to abrasion can be achieved by the incorporation of large amounts of fairly large-particle-size inorganic filler pigments, e.g., $CaCO_3$. The rough paint film that results scatters light and has a "flat" finish, a characteristic of all early latex paints. By incorporating a low-volatility leveling solvent, latex "enamels" (they are not true enamels in the traditional sense) are produced. The polymer is formulated to have a T_g *above* use temperature. It is plasticized by the solvent so that its T_g is *below* application temperature, allowing the particles to coalesce to form a film. Over a period of a day or so after application, the solvent evaporates, leaving the hard, scrub-resistant, high-T_g polymer film, with a reasonable gloss if desired.

Early latex paints were based on styrene–butadiene copolymers or polyvinyl acetate. These have been largely supplanted by paints based on acrylic (acrylates, $ROOCH=CH_2$, or methacrylates, $ROOCCH_3=CH_2$) latices, because of the acrylics' superior chemical stability and, therefore, resistance to color change and degradation.

Much work has gone into adjusting the formulations of latex paints to allow high-pigment contents and thick films for one-coat coverage, together with reasonable ease of application. The main rheological property desired is *thixotropy* (Chapter 15), which gives a high viscosity to prevent the settling of pigments and sagging and dripping of the applied film, together with a lower viscosity under the shearing of brushing or rolling for easy application. These properties are achieved by the addition of finely divided inorganics and water-soluble polymers.

23.3   SOLVENTLESS COATINGS

Traditionally, the application of most surface finishes has been accompanied by the evaporation of an organic solvent. These evaporated solvents are now recognized as a significant source of air pollution. Moreover, since the hydrocarbon crunch, it has become economically unattractive simply to lose them to the atmosphere. While the water-based latex paints have gone a long way toward eliminating these problems, the coatings produced are sometimes lacking in thickness, uniformity, and durability. As a result, there is considerable interest in solventless or *100% solids* coatings.

Two techniques for producing surface coatings without solvent evaporation, plastisol coating and fluidized-bed coating, have been discussed in Section 19.5. Like the latter, *electrostatic spraying* makes use of a polymer powder. The powder is sprayed past an electrode, charging the particles, and onto a heated substrate with the opposite charge. The powder contacting the substrate melts to form a continuous film. Thermosets may be further oven-cured, if necessary. The resulting film is usually quite uniform. Another advantage of this process over traditional spraying of either latex- or solvent-based paints is that the material that is sprayed but not actually deposited on the surface (*overspray*) can be recovered for reuse. Since overspray may account for most of the material sprayed, this can be a significant economy.

The other approach to 100%-solids coatings is to carry out the polymerization reaction right on the surface. Reactive liquid monomer or oligomer (low-molecular weight polymer) is deposited on the surface and polymerized there. The reactants may be two components of a condensation pair (e.g., epoxies – Chapter 2, Example 4, O; or polyurethanes – Chapter 2, Example 4, Q) which are mixed just prior to application and heat-cured on the surface, or they may be unsaturated materials that undergo addition polymerization (acrylates, $ROOC-CH=CH_2$, are favored here because of their high reactivity).

In *radiation curing* processes, addition reactions are activated by various forms of radiation: infrared, microwave, radio frequency, ultraviolet, and electron beam.[1] Basically, the energies of the first three are such that they simply thermally activate the system, that is, heat it, and they are used in conjunction with the usual free-radical initiators. Ultraviolet-cured systems are achieving significant commercial importance, not only as coatings, but also as printing inks, as a result of their low energy requirements (as compared to heat-cured systems) and freedom from pollution. The polymeric portion of a typical ultraviolet-cured coating has three main ingredients: (a) a reactive oligomer, (b) a multifunctional acrylate monomer, and (c) a photoinitiator. The reactive oligomer can be just about any low-molecular-weight polymer containing at least a couple of double bonds, preferably acrylate double bonds because of their high reactivity. A simple example would be the system formed by reacting an excess of 1,6 hexanediol, $HO-(CH_2)_6OH$, with adipic acid, $HOOC-(CH_2)_4COOH$. The resulting hydroxyl-capped polyester is then condensed with acrylic acid, $HOOC-CH=CH_2$, to give the diacrylate polyester

This material, being tetrafunctional, will cure in an addition reaction, but the crosslink density tends to be too low, and even though x is kept low, the viscosity too high for most applications; so it is diluted with a multifunctional acrylate monomer, for example, trimethylol propane triacrylate,

This reduces the viscosity of the mix for easier application and increases the crosslink density in the cured film. The film is cured by free radicals generated by the ultraviolet-induced decomposition of a photoinitiator, for example, benzoin methyl ether,

Electron-beam curing systems are similar in composition. Although they have been around a long time, they have not yet achieved widespread commercial application, probably because of the high capital investment required to produce the electrons and the need to shield the equipment (stray x-rays are generated). Nevertheless, for highly pigmented systems, where ultraviolet radiation tends to be absorbed preferentially near the surface, they offer more uniform cures, and the economics appear reasonable in high-volume applications; so as equipment continues to improve, they should increase in importance in the future.

The chemistry of polymers for high-solids coatings has been extensively reviewed.[2]

23.4   ELECTRODEPOSITION[3]

The process of electrodeposition has been developed to provide uniform, highly adhesive, corrosion-resistant primer coats, particularly for the automotive industry. (The ultimate solution to the corrosion problem is to replace the metal with plastic, but that's another story.) In the most common form of the process, polymers that contain carboxylic acid functionality are produced. These polymers are then "solubilized" in an aqueous medium by partial neutralization with a base to give macroanions (P represents a polymer backbone)

$$P \text{--}(\overset{O}{\overset{\|}{C}}\text{-OH})_n \;+\; m\,BOH \;\longrightarrow\; (HO\text{-}\overset{O}{\overset{\|}{C}}\text{----}P\text{--}(\overset{O}{\overset{\|}{C}}\text{-O}^-)_m \;+\; m\,B^+ \;+\; m\,H_2O$$

Macrocarboxylic   Base   Macroanion
acid

The macroanions are believed to exist in solution as micelles.

The part to be coated is made an anode (+) and immersed in a tank containing the macroanion solution plus pigment, plasticizer, curing agent, and other ingredients. The applied electric field draws the macroanions to the surface where they lose their charge and are deposited as a film. From the standpoint of forming a continuous, uniform, and therefore highly corrosion-resistant film, this process has significant advantages. The electric field draws the articles into all the nooks and crannies of the part (good "throwing power"). Sharp exterior corners, which in other processes tend to be thinly coated because surface tension draws the material away from the corner, produce a stronger electric field and thus attract more polymer. As the deposited coating is a dielectric, the field is stronger at holes in the film, preferentially attracting polymer to seal the hole. Furthermore, pollution and fire hazards are minimized, and the process wastes very little material (unlike spraying) and so offers economic advantages.

A cationic deposition process has recently been introduced commercially.

Theoretically, it offers the best potential for rust resistance. This process is based on amine-functional polymers treated with an acid.

$$P \text{---} \left(NR_2 \right)_n \ + \ mHX \ \longrightarrow \ \left(R_2N \text{---}_{n-m} P \text{---} \left(NR_2H^+ \right)_m \right) \ + \ m\,X^-$$

Acid   Macrocation

The macrocations are then deposited on a cathode.

Following electrodeposition, objects are usually rinsed to remove any non-adhering material and then baked to dry and fuse the film, and perhaps cross-link it.

REFERENCES

1. G. W. Gruber, *Applied Polymer Science*, J. K. Craver and R. W. Tess, Eds., Organic Coatings and Plastics Chemistry Division of the American Chemical Society, Washington, 1975, Chapter 24.

2. R. D. Athey, Jr., *Prog. Org. Coatings*, 7, 289 (1979).

3. F. Beck, *Prog. Org. Coatings*, 4, 1 (1976).

CHAPTER 24

Adhesives

Adhesives have been a technologically important application of polymers for thousands of years. Many of the early natural adhesives such as starch and those based on proteins, for example, hydrolyzed collagen from animal hides, hooves and bones and casein from milk, are still important commercially, but as new adhesive formulations based on synthetic polymers (often the same polymers used in other applications) continue to be developed, the range of applications for adhesives has expanded dramatically.[1,2]

An adhesive has been defined as a substance capable of holding materials (*adherends*) together by surface attachment. Adhesives offer a number of significant advantages as a means of bonding: (a) They are often the only practical means available, particularly in the case of small adherends. For example, it's hard to imagine welding abrasive grains to a paper backing to make sandpaper or bolting the grains together to make a grinding wheel. (b) In the adhesive joining of large adherends, forces are fairly uniformly distributed over large areas of the adherend, resulting in low stresses, and holes (necessary for riveting or bolting) that invariably act as stress concentrators in the adherends are not necessary, thus lowering the possibility of adherend failure. (c) In addition to joining, adhesives may also act as seals against the penetration of fluids. In the case of corrosive fluids, this, coupled with the absence of holes, where corrosion usually gains an initial foothold, can minimize corrosion problems. (d) In terms of weight, it doesn't take much adhesive to join much larger adherends. Hence it is not surprising that many of the newer high-performance adhesives were originally developed for aerospace applications. (e) Adhesive joining may offer economic advantages, often by reducing the hand labor necessary for other bonding techniques.

A detailed treatment of the science of adhesion is beyond the scope of this chapter. Nevertheless, some important generalizations are drawn. Adhesion results from (a) mechanical bonding between the adhesive and adherend, and

(b) chemical forces — either primary covalent bonds or polar secondary forces between the two. The latter are thought to be the more important, and this explains in part why inert, nonpolar polymeric substrates, for example, polyethylene and polytetrafluoroethylene, are very difficult to adhesive bond — they must first be chemically treated to introduce polar sites on the surface. To promote the former, adherend surfaces are often roughened before joining, but this procedure is sometimes counterproductive. It can trap air bubbles at the bottom of crevices, which act as stress concentrators that promote failure in rigid adhesives.

With good bonding between adhesive and adherend, joint failure is cohesive (the adhesive itself fails). Under these conditions, to a good approximation, the properties of the adhesive polymer determine the properties of the adhesive joint; that is, the bond can be no stronger than the glue line. Brittle polymers give brittle joints, polymers with high shear strengths give bonds of high shear strength, heat-resistant polymers produce bonds with good heat resistance, and so forth.

To form a successful joint, the adhesive must intimately contact the adherend surface. This requires first that it *wet* the surface. The subject of wetting is considered in detail in treatises on surface chemistry.[3-8] In general, wetting is promoted by polar secondary forces between adhesive and substrate. This is another reason why low-polarity polymeric adherends such as polyethylene and polytetrafluoroethylene are difficult to adhesive bond. In order to ensure proper wetting and interfacial bonding, it is often necessary to clean the adherend surfaces carefully before joining. Good contact also requires a viscosity low enough under conditions of application to allow the adhesive to flow over the surface and into its nooks and crannies. Once contact has been established, the adhesive must harden to provide the necessary joint strength. There are five general categories of organic adhesive that accomplish these objectives in different ways.

1. Solvent-based Adhesives.   Here the adhesive polymer is made to flow by dissolving it in an appropriate solvent to form a *cement*. The adhesive hardens by evaporation of the solvent. Thus, the polymers used must be linear or branched to allow solution, and the joints formed will not be resistant to solvents of the type used initially to dissolve the polymer. To get a good bond, it helps if the solvent attacks the adherend also. In fact, solvent alone is often used to "solvent weld" polymers, dissolving some of the adherend to form an adhesive on application. One of the drawbacks to solvent-based adhesives based on rigid polymers is the shrinkage that results when the solvent evaporates. This can set up stresses that weaken the joint. An example of this type of adhesive is the familiar model airplane cement, basically a cellulose nitrate solution, with perhaps some plasticizer. Rubber cements, of course, maintain their flexibility, but cannot support as great a stress. Commercial rubber cements are based on natural, SBR (poly(butadiene-co-styrene)), nitrile (poly(butadiene-co-acrylo-nitrile)), chloroprene (poly(2-chlorobutadiene)), and reclaimed (devulcanized) rubbers. Examples are household rubber cement and Pliobond®. Rubber cements may also incorporate a curing agent to crosslink the polymer after application and evaporation of the solvent. This greatly increases solvent resistance and strength.

   2. **Latex Adhesives.**   Latex adhesives are based on polymer latices made by emulsion polymerization. They flow easily while the continuous water phase is present and dry by evaporation of the water, leaving behind a layer of polymer. In order that the polymer particles coalesce to form a continuous joint and that they be able to flow to contact the adherend surfaces, the polymers used must be above their glass-transition temperature at use temperature. These require-ments are similar to those for latex paints; so it is not surprising that some of the same polymers are used in both applications, for example, styrene-butadiene copolymers and polyvinyl acetate. Nitrile and neoprene rubbers are used for increased polarity. A familiar example of a latex adhesive is "white glue," basically a plasticized polyvinyl acetate latex. Latex adhesives are displacing solvent-based adhesives in many applications because of their reduced pollution and fire hazards. They are used extensively for bonding pile to backing in carpets.

   3. **Pressure-Sensitive Adhesives.**   Pressure-sensitive adhesives are really viscous polymer melts at room temperature; so the polymers used must be above their glass transitions. They are caused to flow and contact the adherends by applied pressure, and when the pressure is released, the viscosity is high enough to withstand the stresses produced by the adherends, which obviously cannot be very great. The key property for a polymer used in this application is *tack*, which basically is a viscosity low enough to permit good surface contact, yet high enough to resist separation under stress — something on the order of $10^4 - 10^6$ centipose[1] — although elasticity probably plays a role also. Natural, SBR, and reclaimed rubbers are common in this application. The many varieties of pressure-sensitive tape are faced with this type of adhesive.

   Contact cements are a variation in which the rubbery polymer is applied to *each* adherend surface in the form of a solution or, increasingly, a latex. Evapor-ation of the solvent or water leaves a polymer film with the tack necessary to grab and hold the adherends when they are pressed together.

   4. **Hot-Melt Adhesives.**   Thermoplastics often form good adhesives simply by being melted to cause flow and then solidifying on cooling after contacting the surfaces under moderate pressure. Polyamides and poly(ethylene-co-vinyl acetate) are used frequently as hot-melt adhesives. Electric "glue guns" that operate on this principle have been introduced to the consumer market.

   5. **Reactive Adhesives.**   Reactive adhesives are either monomers or

low-molecular-weight polymers that solidify by a polymerization and/or cross-linking reaction after application. They can develop tremendous bond strengths, have good solvent resistance, and good (for polymers, anyhow) high-temperature properties. The most familiar example of reactive adhesives are the epoxies (Chapter 2, Example 4, O) generally cured by multifunctional amines. Poly-urethanes (Chapter 2, Example 4, Q) also make excellent reactive adhesives. The α-alkyl cyanoacrylate "super glues" ("one drop holds 5000 lb") have recently been introduced to the consumer market. The monomers have extremely low viscosity and so can crawl into narrow crevices and wet the adherend surfaces rapidly. On the other hand, they won't fill gaps and are absorbed into porous adherends. They polymerize in seconds by an anionic addition reaction believed initiated by hydroxyl ions from water adsorbed on the adherend surfaces

Unfortunately (or fortunately, if you stick your fingers together), being linear, they have poor resistance to polar solvents (acetone is a good solvent), and they are subject to hydrolysis; thus they have poor environmental resistance.

Phenolic and other formaldehyde condensation polymers are also important reactive adhesives. Powdered phenolic resin is mixed with abrasive grains, and the mixture is compression molded to form grinding wheels. A B-stage phenolic (Chapter 20) in a solvent is used to impregnate tissue paper. The solvent is evaporated, and the dry sheets are placed between layers of wood in a heated press where the resin first melts and then cures, bonding the wood to form plywood. Unlike the previous examples of reactive adhesives, the phenolics and other formaldehyde condensation polymers evolve water as they cure. If this water is trapped in the joint, serious weakness can result; thus their adhesive applications are limited.

Note that all these examples of reactive adhesives are highly polar polymers. It is largely this polarity that accounts for their good bonding capabilities.

REFERENCES

1.   I. Skeist, "Adhesive Compositions," *Encyclopedia of Polymer Science and Technology*, Vol. 1, N. Bikales, Ed., Wiley, New York, 1971, p. 482.

2. R. L. Patrick, "Chemistry and Technology of Adhesives," *Applied Polymer Science*, J. K. Craver and R. W. Tess, Eds., American Chemical Society, Washington, 1975, Chapter 34.

3. W. A. Zisman, "Effect of Chemical Constitution," *Encyclopedia of Polymer Science and Technology*, Vol. 1, N. Bikales, Ed., Wiley, New York, 1971, p. 450.

4. H. Schonhorn, "Surface Properties," *Encyclopedia of Polymer Science and Technology*, Vol. 13, N. Bikales, Ed., Wiley, New York, p. 535.

5. H. Schonhorn, *Wetting*, SCI Monograph No. 25, Society of the Chemical Industry, London, 1967.

6. F. M. Fowkes, Ed., "Contact Angles, Wettability and Adhesion," *Adv. Chem.*, **43**, American Chemical Society, Washington, 1964.

7. A. W. Adamson, *Physical Chemistry of Surfaces*, 3rd ed., Wiley, New York, 1976.

8. P. C. Hiemenz, *Principles of Colloid and Surface Chemistry*, Marcel Dekker, New York, 1977.

Selected Readings

Aklonis, J., W. J. MacKnight, and M. Shen, *Introduction to Polymer Viscoelasticity,* Wiley, New York, 1971.

Albright, L. F., *Processes for Major Addition-Type Plastics and Their Monomers,* McGraw-Hill, New York, 1974.

Alfrey, T., and E. F. Gurney, *Organic Polymers,* Prentice-Hall, Englewood Cliffs, N.J., 1967.

Allen, P. E. M., and C. E. Patrick, *Kinetics and Mechanisms of Polymerization Reactions,* Wiley, New York, 1974.

Baer, E., Ed., *Engineering Design for Plastics,* Reinhold, New York, 1964.

Billmeyer, F. W., Jr., *Textbook of Polymer Science,* 2nd ed., Wiley, New York, 1971.

Bird, R. B., et al, *Dynamics of Polymeric Liquids,* Vol. 1: Fluid Mechanics, Vol. 2: Kinetic Theory, Wiley, New York, 1977.

Boenig, H. V., *Structure and Properties of Polymers,* Wiley, New York, 1973.

Bovey, F. A., and F. H. Winslow, Eds., *Macromolecules. An Introduction to Polymer Science,* Academic, New York, 1979.

Brandrup, J., and E. H. Immergut, Eds., *Polymer Handbook,* 2nd Ed., Wiley, New York, 1975.

Braun, D., H. Cherdron, and W. Kern, *Techniques of Polymer Synthesis and Characterization,* Wiley, New York, 1972.

Brydson, J. A., *Plastics Materials,* Van Nostrand, Princeton, N.J., 1966.

Chien, J. C. W., Ed., *Coordination Polymerization,* Academic, New York, 1975.

Coleman, B. D., H. Markovitz, and W. Noll, *Viscometric Flows of Non-Newtonian Fluids,* Springer-Verlag, New York, 1966.

Craver, J. K., and R. W. Tess, Eds., *Applied Polymer Science,* American Chemical Society, Washington, 1975.

Driver, W. E., *Plastics Chemistry and Technology,* Van Nostrand Reinhold, New York, 1979.

Elias, H. G., *Macromolecules,* Vol. 1: Structure and Properties, Vol. 2: Synthesis and Materials, Plenum Press, New York, 1977.

Ferry, J. D., *Viscoelastic Properties of Polymers,* 3rd ed., Wiley, New York, 1980.

Flory, P. J., *Principles of Polymer Chemistry,* Cornell University Press, Ithaca, New York, 1953.

Gait, A. J., and E. G. Hancock, *Plastics and Synthetic Rubbers,* Pergamon Press, Elmsford, New York, 1970.

Gorden, M., *High Polymers,* Addison-Wesley, Reading, Mass., 1963.

Harris, F. W., and R. B. Seymour, Eds., *Structure – Solubility Relationships in Polymers,* Academic, New York, 1977.

Kaufman, H. S., and J. J. Falcetta, Eds., *Introduction to Polymer Science and Technology,* Wiley-Interscience, New York, 1977.

Lenz, R. W., *Organic Chemistry of Synthetic High Polymers,* Interscience, New York, 1967.

McKelvey, J. M., *Polymer Processing,* Wiley, New York, 1962.

Meares, P., *Polymers: Structure and Bulk Properties,* Van Nostrand, Princeton, N.J., 1965.

Middleman, S., *The Flow of High Polymers,* Wiley, New York, 1968.

Middleman, S., *Fundamentals of Polymer Processing,* McGraw-Hill, New York, 1977.

Miles, D. C., and J. H. Briston, *Polymer Technology,* Chemical Publishing, New York, 1965.

Miller, M. L., *The Structure of Polymers,* Reinhold, New York, 1966.

Nielsen, L. E., *Mechanical Properties of Polymers,* Marcel Dekker, New York, 1974.

Odian, G., *Principles of Polymerization,* McGraw-Hill, New York, 1970.

Rodriguez, F., *Principles of Polymer Systems,* 2nd ed., McGraw-Hill, New York, 1981.

Saunders, K. J., *Organic Polymer Chemistry,* Chapman and Hall, London, 1970.

Schildknecht, C. E., Ed., *Polymer Processes,* Interscience, New York, 1956.

Schildknecht, C. E., and I. Skeist, Eds., *Polymerization Processes,* Wiley-Interscience, New York, 1977.

Schultz, J. M., *Polymer Material Science,* Prentice-Hall, Englewood Cliffs, N.J., 1974.

Seymour, R. B., *Modern Plastics Technology,* Reston Publishing, Reston, Va., 1975.

Skelland, A. H. P., *Non-Newtonian Flow and Heat Transfer,* Wiley, New York, 1967.

Sorenson, W. R., and T. W. Campbell, *Preparative Methods of Polymer Chemistry,* 2nd Ed., Interscience, New York, 1968.

Stevens, M. P., *Polymer Chemistry,* Addison-Wesley, Reading, Mass., 1975.

Tadmor, Z., and C. G. Gogos, *Principles of Polymer Processing,* Wiley, New York, 1979.

Throne, J. L., *Plastics Process Engineering,* Marcel Dekker, New York, 1979.

Tobolsky, A. V., *Properties and Structure of Polymers,* Wiley, New York, 1960.

Tobolsky, A. V., and R. F. Mark, Eds., *Polymer Science and Materials,* Wiley, New York, 1971.

Treloar, L. R. G., *Introduction to Polymer Science,* Springer-Verlag, New York, 1970.

Tsuruta, T., and K. E. O'Driscoll, Eds., *Structure and Mechanism in Vinyl Polymerization,* Marcel Dekker, New York, 1969.

Van Krevelen, D. W., *Properties of Polymers,* Elsevier, New York, 1976.

Walters, K., *Rheometry,* Halsted Press, New York, 1976.

Ward, I. M., *Mechanical Properties of Solid Polymers,* Wiley, London, 1971.

Williams, D. J., *Polymer Science and Engineering,* Prentice-Hall, Englewood Cliffs, N.J., 1971.

Williams, H. L., *Polymer Engineering,* Elsevier, New York, 1975.

Index